RADIATION EFFECTS IN SEMICONDUCTORS AND SEMICONDUCTOR DEVICES

by V. S. VAVILOV and N. A. UKHIN

TRANSLATED FROM RUSSIAN

CONSULTANTS BUREAU · NEW YORK AND LONDON

ISBN 978-1-4684-9071-8   ISBN 978-1-4684-9069-5 (eBook)
DOI 10.1007/978-1-4684-9069-5
Library of Congress Catalog Card Number 76-17409

A Division of Plenum Publishing Corporation
227 West 17th Street, New York, N.Y. 10011

Translated and printed by
FREUND PUBLISHING HOUSE LTD.
P.O.B. 35010, TEL-AVIV, ISRAEL

CONTENTS

CONTENTS cont.

CONTENTS cont.

PART ONE
CHAPTER I
THE THEORY OF THE FORMATION AND NATURE OF RADIATION-INDUCED DEFECTS

The term 'radiation-induced defects' refers to relatively stable lattice imperfections, produced by high-energy particles. Since single crystals are perfect bodies, their electrical and optical properties are strongly influenced by radiation-induced defects. Extensive studies on the influence of high-energy radiation on semiconductors have been carried out during the last fifteen years. Many of the investigations are concerned not only with general problems of semiconductor physics, but also with practical problems. An example of this is the study of the radiation stability of solar silicon cells used in space explorations.

In this book we shall discuss briefly the formation of radiation-induced defects in semiconductors and the experimental methods used to determine the threshold energies of defect formation. We shall also illustrate the nature of energy spectra related to defects in the forbidden band and the nature and stability of defects.

The problem of radiation-induced defects in semiconductors has been the subject of several books and review articles /1,11-15/.

1. Production of Radiation Defects by Fast Electrons and γ - Rays

The theory of radiation damage is based on the assumption that the simplest defect arising in a crystal is a vacancy or a displaced atom with a more-or-less stable position in the interstice. Such imperfections are called Frenkel defects or Frenkel pairs. Frenkel defects are the simplest point defects in a lattice and differ from dislocations and from the more complex and extended imperfections.

Another assumption is based on the existence of a threshold energy usually designated as E_d which must be imparted to an atom in order to displace it into an interstice. Taking into account the "impact" nature of radiation-induced defects F. Seitz assumed that E_d could be several times greater than the energy of adiabatic displacement of atoms from their lattice sites into interstices (25 eV for crystals with an atomic binding energy of about 10 eV).

A knowledge of the symmetry and of atomic binding energies in a lattice can be very helpful in the determination of the mechanism of production of Frenkel defects by fast electrons.

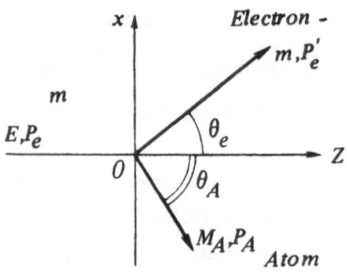

Figure 1. Scattering of fast electrons on atoms.

Let a fast electron with an energy E and a momentum P_e move along the z axis (Figure 1) and collide with an atom at rest. After the collision the electron has a momentum P'_e and continues to move in the plane xoz at an angle θ_e to the z axis. As a result of the collision, the atom has acquired a momentum P_A and it starts to move in the plane xoz at an angle θ_A to the z axis. From the law of conservation of momentum it follows that

$$P_e = P'_e \cos \theta_e + P_A \cos \theta_A;$$
$$0 = P'_e \sin \theta_e - P_A \sin \theta_A;$$
$$P_A^2 = (P_e - P'_e \cos \theta_e)^2 + P'^2_e \sin^2 \theta_e; \qquad (1)$$
$$\text{tg } \theta_A = \frac{P'_e \sin \theta_e}{P_e - P'_e \cos \theta_e}.$$

The mass of the atom M_A is much larger than that of the electron m (e.g. $M_{Si} \simeq 5 \times 10^4 \, m$; $M_{Ge} \simeq 1.3 \times 10^5 \, m$). Thus, $P'_e \simeq P_e$; hence

$$P_A^2 = P_e^2 \sin^2 \left(\frac{\theta_e}{2} \right); \qquad \theta_A = \frac{\pi}{2} - \frac{\theta_e}{2}. \qquad (2)$$

If the kinetic energy of an electron is equal to E and the kinetic energy of an atom after collision is $E_A = P^2/2M_A$, we could write $mc^2 = 0.511 \text{ MeV} = E'_0$ (electron rest energy); from there we have

$$E_A = \varepsilon' \frac{2m}{M_A} \left[\left(\frac{E'}{E_0'}\right)^2 + 2\frac{E}{E_0'} \right] \cos^2 \theta_A = E_{Amax}.\cos^2 \theta_A \quad (3)$$

or

$$\frac{561}{A} \varepsilon(\varepsilon+2)\cos^2 \theta_A \text{ eV},\quad (3a)$$

where A is the atomic weight and $\varepsilon = E/mc^2$.

The maximum energies transferred to atoms of germanium and silicon by electrons of different energies are listed in Table 1 $(\theta_A=0; \theta_e=\pi)$.

Table 1. Maximum energies imparted to germanium and silicon atoms as a function of electron energy.

E_e, MeV	$E_{Ge\ max}$, eV	$E_{Si\ max}$, eV	E_e, MeV	$E_{Ge\ max}$, eV	$E_{Si\ max}$, eV
0.01	0.3	0.8	1.0	59	125
0.1	3.2	8.2	2.0	177	455
0.3	12	30.7	3.0	354	920
0.5	23	59			

Let $\sigma(\theta_A)$ be the cross section which characterizes the probability that the momentum acquired by an atom is at an angle θ_A to the z axis.

Let $\sigma_e(\theta_e)$ be the differential scattering cross section of an electron at an angle θ_e. Then

$$\sigma_e(\theta_e)\sin\theta_e\,d\theta_e = \sigma(\theta_A)\sin\theta_A\,d\theta_A, \quad (4)$$

where θ_e and θ_A are related as follows: $\theta_A = \frac{\pi}{2} - \frac{\theta_e}{2}$.

Equations (2) and (4) indicate that if $\theta_A < \frac{\pi}{2}$ then

$$\sigma(\theta_A) = 4\sigma_e(\pi - 2\theta_A)\cos\theta_A, \quad (5)$$

while if $\theta_A > \frac{\pi}{2}$ we would have

$$\sigma\,(\theta_A) = 0. \tag{6}$$

The shielding of atomic nuclei by electron clouds does not usually affect the collisions between fast electrons and atomic nuclei, which are accompanied by displacement of atoms from their lattice sites /11/. Such collisions can be considered as the scattering of relativistic electrons in the coulomb field. Using an approximation, which is satisfactory for small θ_e and small energies transferred to the atoms, we can take for θ_e the value of the relativistic Rutherford cross section:

$$\sigma_R\,(\theta_e) = \sigma_0 \frac{1}{\sin^4 \theta_c /2}\,,$$

where

$$\sigma_0 = \left(\frac{Ze^2}{2mc^2}\right)^2 \frac{1 - \beta^2}{\beta^4}\,, \tag{7}$$

Z is the charge of the nucleus and $\beta = \dfrac{v}{c}$.

For the Rutherford scattering cross section

$$\sigma\,(\theta_A) = 4\sigma \frac{1}{\text{\i s}^3 \theta_A} \quad \text{at} \quad \theta_A < \frac{\pi}{2}$$

and

$$\sigma\,(\theta_A) = 0 \quad \text{t} \quad \theta_A > \frac{\pi}{2}. \tag{8}$$

The cross section which characterizes the probability of the momentum being directed at an angle θ_A greatly increases as θ_A approaches $\frac{\pi}{2}$. In such a case, the so-called impact parameter is large, and both the angle of electron scattering and the energy transferred are small. Actually, the electron scattering cross section $\sigma_e(\theta_e)$ greatly differs from the value obtained from the Rutherford formula. According to McKinley and Feshbach /16/ for light elements $\sigma_e(\theta_e)$ can be calculated with the aid of the following expression:

$$\sigma_e\,(\theta_e) = \sigma_R\,(\theta_e)\,B \sin \frac{\theta_e}{2}\,; \tag{9}$$

B is given by

$$B\,(x) = 1 - \beta^2 x^2 + \pi\alpha\beta x\,(1 - x), \tag{10}$$

where $x = \sin \theta_e / 2$, and $\alpha = Ze^2/hc$. The value of $\sigma(\theta_A)$ can be obtained from the equation

$$\sigma(\theta_A) = 4\sigma_0 \frac{B(\cos \theta_A)}{\cos^3 \theta_A} \, . \tag{11}$$

The most important conclusion that must be drawn from both approximate and exact measurements is that the momentum transfer is usually at a large angle to the direction of the incident electrons.

The energy spectrum of atoms excited by the scattering of fast electrons can be calculated in the following way. Let $n(E_A)dE_A$ be the number of atoms with an energy between E_A and $E_A + dE_A$ (for a uniform flux of fast electrons). In such a case:

$$n(E_A) dE_A = 2\pi\sigma_e(\theta_e) \sin \theta_e \, d\theta_e \, . \tag{12}$$

From the relationship between the angles θ_e and θ_A we find that

$$E_A = E_{A\ max} \cos^2 \theta_A = E_{A\ max} \sin^2 \frac{\theta_e}{2} \, ;$$

$$dE_A = \frac{1}{2} E_{A\ max} \sin \theta_A \, d\theta_A \, . \tag{13}$$

We can now derive an expression for the relationship between the energy spectrum and the differential scattering cross section of electrons:

$$n(E_A) = \frac{4\pi}{E_{A\ max}} \sigma_e 2 \sin^{-1}\left[\left(\frac{E_A}{E_{A\ max}} \right)^{1/2} \right] . \tag{14}$$

From the Rutherford equation we find that

$$n(E_A) = \frac{4\pi\sigma_0}{E_{A\ max}} \left(\frac{E_{A\ max}}{E_A} \right)^2 \tag{15}$$

Equation (15) shows that most atoms acquire small energies. The relation does not change if we use a more exact formula for the calculation of the scattering cross section. Figure 2 shows the distribution of energy transferred to germanium atoms by electrons with an initial energy of 1.5 or 3 MeV. Except for the sections corresponding to head-on collisions, the curves coincide. Thus, in the irradiation of crystals with electrons of various energies, we could expect that most defects would be produced as a result of collisions which impart to the atoms an energy only slightly greater than the threshold energy.

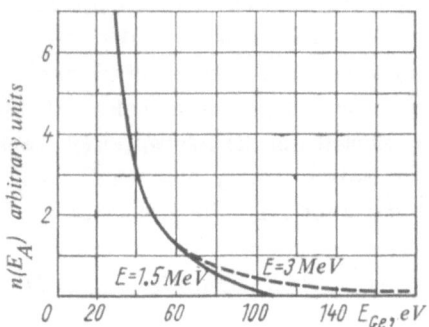

Figure 2. Distribution of energy imparted to germanium atoms by electrons with an initial energy of 1.5 or 3 MeV.

If we assume that the threshold energy E_d, required to displace an atom into an interstice is not affected by the direction of the momentum of the atom P_A and that each scattering which is accompanied by an energy transfer $E_A > E_d$ produces a simple Frenkel defect, then the cross section Σ_d (which characterizes the probability of defect formation) can be calculated from the above equations. Since the probability of defect production depends on the electron energy E, the following equation is valid only if the energy lost by an electron in the crystal (ΔE) is small $(\Delta E \ll E)$, i.e. for thin specimens. The following equation can be used to calculate the concentration of defects N_d produced by an integrated electron flux Φ:

$$N_d = \Sigma_d \Phi N, \tag{16}$$

where N is the number of atoms in $1\,cm^3$.

Obviously, defects should be produced by all collisions, for which the angle θ_A has a value between $\theta_A = 0$ (head-on collision) and $\theta_{A\,max}$ which can be calculated from the following equation:

$$E_{A\,max} \cos^2 \theta_{A\,max} = E_d,$$

i.e.,
$$\cos\theta_{A\,max.} = \sqrt{E_d/E_{A\,max}}. \qquad (17)$$

The final equation for $\Sigma_d(E)$ would be:

$$\Sigma_d(E) = 8\pi\sigma_0 \left[\frac{1}{2}\left(\frac{1}{x_0^2} - 1\right) + \pi\alpha\beta\left(\frac{1}{x_0} - 1\right) + \right.$$
$$\left. + (\beta^2 + \pi\alpha\beta)\ln x_0 \right], \qquad (18)$$

where

$$\sigma_0 = \left(\frac{Ze^2}{2mc^2}\right)\cdot\frac{1-\beta^2}{\beta^4} \; ; \quad \alpha = \frac{Ze^2}{hc} = \begin{cases} 0{,}23 & \text{for Ge,} \\ 0{,}1 & \text{for Si;} \end{cases}$$

$$x_0 = \cos\theta_{A\,max.} = \sqrt{\frac{E_d}{E_{A\,max.}}} \; ;$$

$$\beta = \frac{\left[\frac{2E}{mc^2} + \left(\frac{E}{mc^2}\right)^2\right]^{\frac{1}{2}}}{1 + \frac{2E}{mc^2} + \left(\frac{E}{mc^2}\right)^2} \; .$$

Thus, the function $\Sigma_d(E)$ for crystals made of identical atoms is determined by E_d. Occasionally it is important to know the total number of atoms displaced by bombardment with fast electrons. Kahn, who assumed that the threshold energy E_d is not affected by the direction of the electrons, has calculated the total number of displaced atoms of silicon and germanium $N^+(E)$ per incident electron /17/; the results are shown in Figure 3.

He also assumed that secondary defects would be produced if the energy of the primary displaced atom E_A exceeds $2E_d$. The total number of defects produced by such a collision would be /18/:

$$\bar{\nu} = E_A/2E_d.$$

In a study of scattering in thin metallic foils, carried out to verify the theory of scattering of relativistic electrons by nuclei, which was developed by Mott, McKinley and Feshbach, it was shown that the experimental data agreed well with the theoretically calculated values, with an accuracy within ± 1%. Hence, the mechanism of primary energy transfer from fast electrons to atoms seems quite clear, in contrast to the mechanisms of subsequent stages involving the creation and stabilization of defects.

13

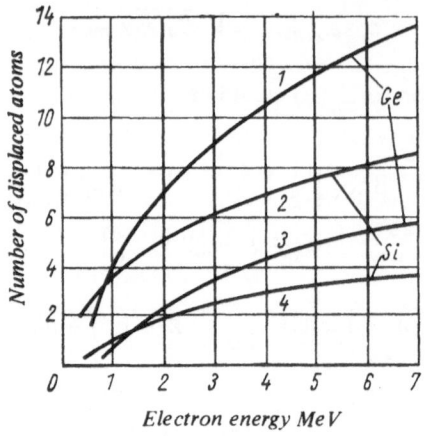

Figure 3. Number of displaced atoms of germanium and silicon per single incident fast electron $[E_d 15 \ /1,2/$ and $30/3,4/$ eV$|$

The forces between an atom to which a momentum has been imparted and the neighboring atoms in the lattice have almost no effect on its initial motion, although in subsequent stages their influence is considerable. The collision time τ_i for electrons with a kinetic energy of several MeV is:

$$\tau_i \simeq \frac{\lambda}{c} \simeq 10^{-20} \ \text{sec.}$$

(where $\lambda = \frac{h}{mc}$ is the Compton wavelength). This is much less than the time necessary to displace an atom through a typical interatomic distance in the crystal; the initial velocity of a displaced atom (if we assume that the entire energy is kinetic) is equal to 10^6 cm/sec. Consequently, a displacement by 10^{-8} cm involves a time of the following order of magnitude:

$$\tau \, displacement \simeq \frac{10^{-8}}{10^6} = 10^{-14} \ \text{sec}$$

2. On the Production of Frenkel Defects in Diamond-type Lattices.

After a collision between an electron and an atom, the binding forces of the displaced atom tend to return it to its initial site. For small displacements these forces can be evaluated using the known elasticity constants. A germanium atom with a kinetic energy of 25 eV has a velocity of the order of 10^6 cm/sec, which is very slow compared to the velocity of electrons of a corresponding binding energy in crystals (about 10^8 cm/sec). Therefore, at each moment the electrons are in a state corresponding to the lowest energy of the given configuration of nuclei, as in the case of elastic deformation of crystals. According to the approximate theory of elasticity constants /19/, the energy ΔE_A necessary to displace an atom in a diamond-type lattice a small distance x in any direction can be found from:

$$\Delta E_A = c' \left(\frac{x}{b} \right)^2 , \qquad (19)$$

where for germanium c' is equal to 18.3 eV and $b = 1.41 \times 10^{-8}$ cm (b is the distance between two neighboring atoms.)

Using the data on "regular" interstice states in diamond-type lattices, equation (19), and known values of the binding energies of atoms in crystals, it is possible to evaluate the loss of energy in the displacement of an atom from its site into the nearest interstice. Obviously, the estimates of energy related to larger distances are not accurate.

Let an atom O, to which kinetic energy was transferred as a result of the scattering of a fast electron, lie at the origin of coordinates. In that case the position of the four closest atoms will be $A \ldots b(1, 1, 1)$; $B \ldots b(\bar{1}, 1, 1)$; $C \ldots b(1, 1, -1)$; and $D \ldots b(1, -1, 1)$ (Figure 4). One of the closest regular interstices O_1 lies on the axis at point $(1, 1, 1)$ at a distance $OO_1 = b\sqrt{3} = 1.73\,b$. The distance from point O to point M at the intersection of the line OO_1 with the plane which contains the three closest atoms is $d = 0.6\,b$. The potential energy $\Delta E_A(OM)$ corresponding to the displacement d_{OM} is equal to about 6 eV. Kohn, who investigated the interaction of an atom O with its neighbors when displaced in the direction OO_1 and in the direction of the closest interstices /20/, found that the energy of displacement of neighboring atoms is smaller than the energy $\Delta E_A(OM)$.

The outer-shell electron configuration of free silicon and germanium atoms is $4s^2 4p^2$. In the crystal lattice the probability of spatial distribution of outer shell electrons corresponds to a combination of the wave functions of the s- and p-states. The maximum probability of occurrence of electrons is along the line connecting neighboring atoms. The energy of valence bonds can be approximately calculated by dividing the binding energy of crystals by four times the number of atoms. Since the experimentally found binding energy of germanium

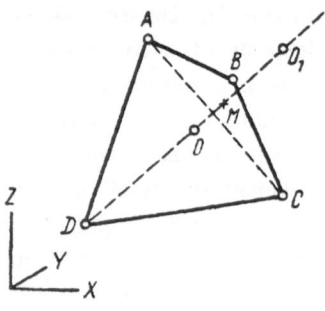

Figure 4. Position of closest neighbouring atoms and of natural interstice states in diamond-type lattices.

crystals is E_{Ge} = 89 cal/mol, the energy of valence bonds will be $E_{Ge}/4N_{Ge}$ = 1.92 eV.

The $0 \rightarrow 0_1$ transition.

As we mentioned above, the displacement of an atom from point 0 to M increases the potential energy by 6 eV and has no influence on cther atoms. The maximum energy E_d for the 00_1 displacement along the (111) axis is the sum of the potential energy $\Delta E_A(OM)$ and of the energy necessary to disrupt four valence bonds; thus,

$$E_d(OO_1) \leqslant (6 + 4 \cdot 1.92) = 13.7\,eV.$$

This value is close to values obtained by Brown and Augustinak (15.3 eV) [21] and by Smirnov (15.7 eV) [23]. However, theoretically, the 00_1 displacement has no potential barrier which could stabilize the newly generated defects.

Displacement into point $b[(2, 0, 0), (0, 2, 0)$ or $(0, 0, 2)]$.

In this case, unlike the case of the 00_1 displacement, the atom O must pass a point between B and C. By using the equation for the calculation of the potential energy of atom displacement and taking into account the interaction between B and C and the rupture of valence bonds, we obtain a higher threshold energy:

$$E_{d_2} = (33.5 \pm 4)\ eV.$$

It should be pointed out that an energy of 3-4 eV is needed for the displacement of atoms B and C. The return of those atoms to positions close to their initial sites can produce a potential barrier which stabilizes the defects.

The other regular interstices around atom O can be occupied by it only after collisions with neighboring atoms, which is accompanied by considerable energy transfer; therefore, the corresponding E_d values should be much larger than those for the displacements we discussed before.

The parameter E_d shows that the threshold energies should indeed be close to those given by Seitz. In the vicinity of the "threshold", characterizing the formation of lattice defects, the number of produced defects is greatly influenced by the mutual orientation of the crystal axis and the direction of the incident electrons.

Gamma-rays, like fast electrons, produce point defects. The probability of an atom being displaced by a direct interaction between γ-rays and atomic nuclei in crystals is very small. Most defects are formed by fast electrons, produced by the photoelectric effect and by the Compton effect, and by pairs of electrons and positrons formed at high γ-ray energies. The total absorption cross section of γ-rays is determined by the above three processes. The number of displaced atoms can be calculated on the basis of the theory of production of Frenkel defects by fast electrons and positrons.

In the case of positrons we must take into account annihilation, which reduces the number of produced defects.

Kahn /17/ calculated the number of silicon and germanium atoms displaced by photons, assuming the existence of threshold energy of defect formation (E_d), at $E_A > E_d$. The results of these calculations for threshold energies of 15 and 30 eV and for γ energies up to 7 MeV are given in Figures 5 and 6.

It has been found experimentally that the spectra of energy levels of defects produced by fast electrons and γ-quanta coincide, which is in agreement with theoretical calculations.

3. Defect Production by Neutrons and by Heavy Charged Particles

Fast neutrons can produce lattice defects and impart part of their kinetic energy to the atomic nuclei. Usually, the recoil nucleus drags with it the electron shell. Only the weakest-bound electrons of the outer shell can conceivably be detached from atoms displaced by neutrons /11, 12/. The most likely process is the elastic scattering of

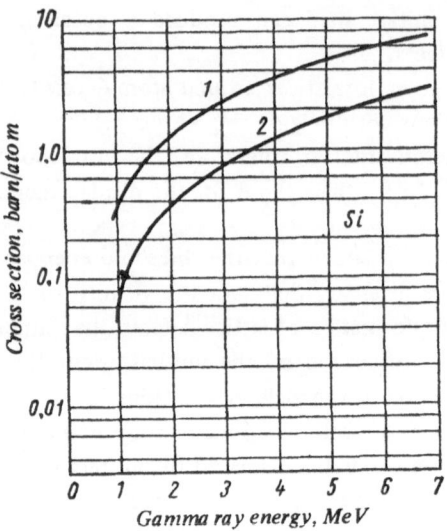

Figure 5. Cross-section σ_d characterizing the probability of displacement of a silicon atom by γ-rays assuming that $E_d = 15$ eV (1) and $E_d = 30$ eV (2).

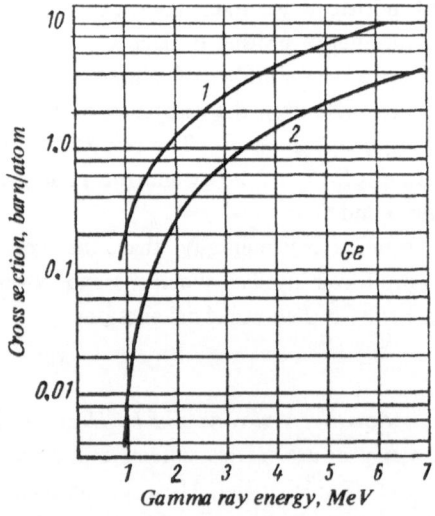

Figure 6. Cross-section σ_d characterizing the probability of displacement of germanium atoms by γ-rays assuming that $E_d = 15$ eV (1) and $E_d = 30$ eV (2).

fast neutrons. The energy transferred to the nuclei can vary from 0 to $E_{A\,max}$:

$$E_{A\,max.} = \frac{4M_n M_A}{(M_n + M_A)^2} E_n, \tag{20}$$

where M_n denotes the neutron mass, and E_n is the kinetic energy of the neutron. The distribution of energy between the recoil atoms is influenced by the angular distribution of the scattered electrons. The simplest assumption is that of isotropic scattering, where all recoil energies from 0 to $E_{A\,max}$ are equally probable. The mean energy transferred by scattering is:

$$E_A = \tfrac{1}{2} E_{A\,max}. \tag{21}$$

For fast neutrons produced by nuclear fission the cross-section which characterizes the probability of scattering is in the range of $(1\text{-}10)\cdot 10^{-24}$ cm^2. Such neutrons have an energy from 0 to 15 MeV and a mean energy of 2 MeV. If we assume that the mean energy of neutrons is equal to 2 MeV* it follows from (20) and (21) that the mean energy of recoil nuclei would be

$$\overline{E}_A = \frac{4}{A} \left(1 + \frac{1}{A}\right)^{-2} \simeq \frac{4}{A} \ \ \text{MeV} \tag{22}$$

where A is the atomic weight.

During fast-neutron bombardment the mean energy of recoil atoms is much greater than the mean energy of recoil atoms produced by bombardment with charged particles (e.g. α-particles, protons) of the same energy.

It has been found experimentally that the assumption of isotropic scattering of fast neutrons is not accurate. In reality, neutrons with an energy of several MeV tend to scatter in a forward direction and therefore the neutron energy transfer is smaller than the value calculated from equation (21). Elastic scattering is also possible and this leads to a decrease in the mean kinetic energy of the recoil atoms.

For neutron energies about 1-2 MeV the anisotropy of scattering reduces the E_A value of most elements by 30-50%. For higher neutron

---

* During irradiation in a reactor the spectrum of neutrons impinging on a specimen can be different from the spectrum of non-interacting neutrons, depending on the conditions of experiment and primarily on the geometry and material of the moderator.

energies the difference is even greater.

We have already said that the kinetic energy transferred by fast neutrons to atomic nuclei during scattering, considerably exceeds the threshold of defect formation E_d. This is also true for bombardment with heavy charged particles having an energy of several MeV. Since the primary displaced atoms can produce secondary displacements, the total number of lattice defects is always greater than the number of scatterings of fast neutrons in the crystal. The number of primary displaced atoms in a unit volume of atoms N_p can be calculated:

$$N_p = \Phi N_A \sigma_d, \tag{23}$$

where Φ is the integrated flux; N_A is the number of atoms in 1 cm³ and σ_d is the collision cross section which leads to primary displaced atoms*.

It is assumed that for neutron bombardment $\sigma_d = \sigma_T$, where σ_T is the total cross section of neutron interactions. Let ν be the mean number of atoms knocked out by a primary atom including the primary atom itself. The value of ν depends on the energy of the primary atoms, the mean value shall be $\bar{\nu}$. The total number of displaced atoms in 1 cm³, N_d can be found from:

$$N_d = \bar{\nu} N_p. \tag{24}$$

Henceforth we shall consider all collisions as paired ones since the radius of action of forces is much smaller than interatomic distances in the crystal. It is also assumed that atoms in the crystal lattice are in a state of rest. Ordering of atoms is not taken into account. Alternate assumptions have been made particularly by Kinchin and Rease /18/ who assumed that:

a. The primary atom loses energy on ionization of the substance, until its kinetic energy reaches a limiting value of E_i which, according to Seitz, can be found from the equation:

$$E_i = \frac{1}{8} \frac{M_i}{m} E_g, \tag{25}$$

where M_i is the mass of an atom in motion; m is the mass of the electron; E_g is the lowest excitation energy of electrons which coincides with the "optical width" of the forbidden gap.

b. All moving atoms, with energies smaller than E_i lose energy only as a

---

* Henceforth we shall use the term "primary" for an atom knocked out of its site, which has a considerable kinetic energy.

result of elastic collisions with atoms in the crystal lattice.

c. An atom leaves its site if the impact of another atom imparts to it a kinetic energy E_A exceeding the threshold energy E_d, and it remains at its site if $E_A < E_d$.

d. An incident atom displaces a knocked-out atom if the latter acquires an energy greater than E_d while the incident atom is left after collision with an energy $E_A < E_d$. Thus, the total number of defects increases only if both atoms after the collision have energies exceeding E_d.

According to the above model the atoms cannot overcome the potential barrier until they collide with other atoms and displace them. The atom leaves its site possessing the entire kinetic energy transferred to it by the collision.

On the other hand Seitz /11/, and Snyder and Neufeld /23/ have assumed that an atom loses part of its kinetic energy (equal to E_d) before it can displace other atoms. It may also be assumed that an incident atom cannot remain at the site of a displaced atom. These different assumptions to a great extent cancel each other in quantitative calculations.

Analyses of the cascade process are usually based on the laws of collision of rigid spheres; it has been found that introduction of Rutherford's law which accurately describes collisions at relatively high energies, does not change significantly the number of displaced atoms.

The dependence of the mean number of defects $\bar{\nu}$ on E_A^* can be calculated as follows. Let E be smaller than E_i but larger than $2E_d$. In an elastic collision the first atoms transfer to the other atoms an energy E_2' and conserve an energy E_1'; $E_1' + E_2' = E_1$. According to the law of collision of rigid spheres all values of transferred energy from 0 to E_i are equally probable and the differential cross section $d\sigma$ for energy transfer in the range $E \ldots E + dE$ is equal to:

$$d\sigma = c' \, dE; \quad c' = \frac{\pi a_1^2}{E_1}, \qquad \cdot(26)$$

where a_1 is the radius of the sphere. After the first exchange of energies the number of atoms that can be displaced by the first atom is equal to $\nu(E_1')$, if $E_1' \geqslant E_d$, or to 0 if $E_1' < E_d$. The mean number of displacements by the first atom can be found from:

---

*Henceforth the subscript A shall be omitted.

$$\int\limits_{E_d}^{E_1} \frac{1}{E_1} \, \nu \, (E_1') \, dE_1'. \qquad (27)$$

The number of displacements by the second atom is equal to $\nu(E_2)$ if $E_2' \geqslant E_d$, and it is equal to 0 if $E_2' < E_d$. By multiplying the number of displacements by the probability of a given energy distribution and integrating with respect to energy we obtain a formula similar to (27).

By summing up both integrals we obtain an equation for $\nu(E)$ which is valid for the range $2E_d \leqslant E < E_i$:

$$\nu \, (E) = \frac{2}{E} \int\limits_{E_d}^{E} \nu \, (E') \, dE'. \qquad (28)$$

If we multiply both sides of the above expression by E and differentiate with respect to E we obtain:

$$E \, \frac{d\nu}{dE} = \nu \quad \text{in the range } 2E_d < E < E_i, \qquad (29)$$

which yields $\nu \, (E) = CE$.

The constant C is determined from the condition $\nu \, (2E_d) = 1$:

$$\left.\begin{aligned}
\nu \, (E) &= 1 && \text{at} && 0 < E < 2E_d, \\
\nu \, (E) &= \frac{E}{2E_d} && \text{at} && 2E_d < E < E_i, \\
\nu \, (E) &= \frac{E_i}{2E_d} && \text{at} && E > E_i.
\end{aligned}\right\} \qquad (30)$$

Thus, on the average, half the energy of primary atoms is used in the formation of defects while the second half is lost in collisions which do not produce any displacements.

A schematic presentation of the $\nu \, (E)$ relationship, plotted according to the Kinchin model (lower curve) and to the Seitz model (top curve) is shown in Figure 7. In the vicinity of $E = E_d$ the difference between these curves is considerable but for high energies these curves are almost identical. The results obtained by Seitz can be approximately represented by the equation:

$$\nu\left(E\right) = 0.56 + 0.56\,\frac{E}{E_d}\,, \tag{31}$$

In order to calculate the total number of defects in a crystal, the value of ν must be averaged over the energy spectrum of the primary atom.

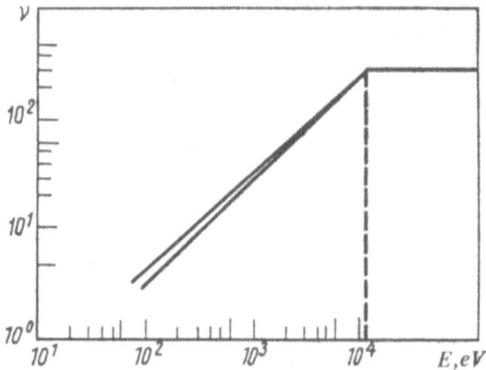

Figure 7. Dependence of the number of displaced atoms, ν, on the energy of the primary atom (cascade theory) of germanium ($E_d = 15$ eV; $E_i = 12{,}000$ eV).

Bombardment of atoms (that are not too light) with fast neutrons in a reactor, imparts to most primary atoms an energy at which $\nu\left(E\right)$ is a linear function of E and therefore E (in $\nu = E/2E_d$) can be replaced by $E = f \cdot \frac{1}{2}\,E_{max}$:

$$\nu = f\,\frac{2M_n M_A}{(M_n + M_A)^2} \cdot \frac{\bar{E}_n}{E_d}\,, \tag{32}$$

where $\bar{E}_n$ is the mean kinetic energy of neutrons and f is a factor for the anisotropy of scattering.

Bombardment or irradiation with high-energy neutrons leads to the loss of a considerable part of the energy of primary atoms on ionization and therefore averaging is more complicated.

During irradiation with heavy charged particles, the energies of the primary atoms are very small and the $\nu\left(E\right)$ relationship can be considered a linear function of E. The value of ν can be found from:

$$\bar{\nu} = \nu\left(\bar{E}\right), \tag{33}$$

where the mean energy transferred by Rutherford scattering is:

$$\bar{E} = \left(\frac{E_d E_{max}}{E_{max} - E_d)} \right) \ln \frac{E_{max}}{E_d} , \tag{34}$$

A more accurate equation has been derived by Dienes:

$$\bar{\nu} = \frac{1}{2} \left(\frac{E_{max}}{E_{max} - E_d} \right) \left(1 + \ln \frac{E_{max}}{2E_d} \right) . \tag{35}$$

Hence, we see that bombardment with 1 MeV charged-particles produces defect clusters much smaller than in the case of bombardment with fast neutrons in a reactor; this is attributed to differences between scattering in the coulomb field and scattering of neutrons by the nuclei.

All cascade models are approximate and the calculated results can greatly differ from the real number of displaced atoms.

In addition to direct displacement into interstices, irradiation of a crystal may lead to displacement of defects as a result of a chain of successive momentum transfers from the moving atoms to the atoms at rest, in their regular lattice sites.*

The theory of the cascade process, based on an analysis of the collision of rigid spheres and of displacements by collision /18/, clearly differentiates between ionization losses and elastic collisions. The authors also made additional assumptions:

a. Atoms leave their regular sites with a kinetic energy greater than E_d. In addition, if the energy transferred to an atom is between E_d and E_r where $E_r < E_d$ and the energy of the displaced atom is smaller than E_d then the striking atom will take the site of the knocked atom, which means that the defect is displaced into a new position.

b. The striking atom replaces the knocked atom if the latter acquires an energy greater than E_d and the former is left, after collision, with an energy less than E_d.

As in the models discussed above, the number of defects can be increased only if both the striking and the knocked atoms have, after collision, an energy greater than E_d. Therefore, the total number of defects remains constant. However, the calculated number of atom displacements in the lattice increases, depending on the threshold

---

* The process of successive transfer of energies and momenta along a chain of atoms is called "focused collision" /24/.

displacement energy E_r. The method of calculation of the number of displacements is similar to that used before for the calculation of the numbers of defects $\nu(E)$. The mean number of displacements $\mu(E)$ produced by a primary atom with a kinetic energy E can be found from the equation:

$$\mu(E) = \frac{E}{2E_d}\left(1.6\ln\frac{E_d}{E_2} + 1\right) ; \quad E \geqslant E_d.$$ (36)

A comparison of the values of $\mu(E)$ and $\nu(E)$ yields:

$$\frac{\mu(E)}{\nu(E)} = 1.6\ln\frac{E_d}{E_r} + 1.$$ (37)

Hence if E_r is much smaller than E_d, each new defect is accompanied by several displacements. No direct evaluations of E_r are available but the authors of /18/ believe that $E_d \cong 10E_r$; hence $\mu(E) \cong 5\nu(E)$.

Much of the energy transferred to one atom in the crystal during the primary act of scattering of a particle (for instance a neutron) is subsequently distributed between a large number of atoms. The state of the substance in the vicinity of the primary energy transfer can be approximately represented as fast heating of a limited volume to a high temperature. In addition to the cascade multiplication of defects, their final number, which determines the properties of the irradiated crystal, is greatly influenced by the rate of equalization of energy in the field of strong excitation. Certain assessments can be made and a qualitative theory of the processes following the energy transfer to the primary atom can be derived on the basis of the theory of thermal conductivity. It should be pointed out, however, that an equilibrium distribution is not established in the region of excitation, and that its state must not be characterized by the temperature.

The time and the distance which characterize the dissipation of the excitation are so small that macroscopic laws of thermal conductivity can be used only for a qualitative description.

Seitz and Keller believe that at the first instant the excitation can be so intense that a large fraction of the atoms in the excited crystals are disordered (displacement spike).

According to the thermal theory, the energy E_A transferred to an atom by a particle is regarded as heat suddenly evolved in a small volume of a continuous medium, its propagation is governed by the classic laws of thermal conductivity. The medium is characterized by a diffusion coefficient D and a temperature $T(r, t)$ at each point r at every instant t. The diffusion coefficient D is related to the thermal conductivity C, the specific heat c and the density d by the equation:

$$D = \frac{C}{cd} . \tag{38}$$

·The thermal conductivity of semiconductors and metals is the sum of terms representing the thermal conductivity of the crystal lattice and the contribution of electrons to the conduction band /7/. This type of excitation in crystals, which is the result of the displacement of atoms in the lattice, does not influence the current carriers, at least not during the crucial initial stage. Thus, the value of C should correspond to the thermal conductivity of the lattice. The diffusion coefficient D is usually close to 10^{-3} cm²/sec. The temperature in the range of excitation is given by the law of thermal conductivity:

$$\nabla^2 T = \frac{1}{D} \cdot \frac{\partial T}{\partial t} . \tag{39}$$

The solution of this equation, which represents the energy evolution from time $t = 0$ (origin of coordinates), at an initial temperature T_o and a distance from the origin of the coordinates r yields:

$$T(r, t) = T_0 + \frac{E_A}{(4\pi)^{3/2} cd} \cdot \frac{1}{(Dt)^{3/2}} e^{\frac{r^2}{4Dt}}. \tag{40}$$

At any time the maximum temperature is in the vicinity of the origin of coordinates and it decreases proportionally to $t^{-3/2}$.

If the energy of the primary atom is high and the distance between consecutive collisions is large enough, then the regions of thermal excitation by collision can be regarded as spherical. As the primary atom slows down, those regions overlap. This process can be considered as a uniform evolution of thermal energy along the track of the primary atom. If the thermal energy, related to a unit length of the particle track, is equal to Q' and the radius is ρ, the solution of the thermal conductivity equation yields:

$$T\,(\rho,\ t) = T_0 + \frac{Q'}{4\pi cd} \cdot \frac{1}{Dt}\ e^{-\frac{\rho^2}{4Dt}}. \tag{41}$$

Brinkman /25/ calculated that the mean distance between collisions accompanied by a displacement of secondary atoms is equal to the interatomic distances if the energy of the primary atom is $2 \cdot 10^4$ ev. Hence it follows that as soon as the energy of the primary atom decreases to that level, the atom slows down rapidly, producing a dense region of secondary displacements so that it is impossible to consider each of the defects in separate. This means that a large number of atoms in a nearly cylindrical volume would be totally disordered as in melts or in vapors.

Brinkman believes that an "inversion" takes place in the vicinity of the track of the primary atom, i.e., that atoms in the vicinity of this track will be displaced to a maximum distance. The disordered region crystallizes again within $10^{-10} - 10^{-12}$ sec., starting from the external boundary. It can be expected that most atoms would be arranged in the same way as in the initial crystal lattice. The final number of defects (vacancies, interstitials and closed-loop dislocations) will not be the same as that calculated on the basis of the cascade theories.

Unfortunately, no definite quantitative results can be obtained on the basis of the Brinkman theory. The basic assumption of this theory (frequency of atom collisions in the region of their elastic interaction, higher than that assumed in other theories) has been criticized. F. Seitz and F. Smirnov believe that the thermal theory of the production of radiation defects can be applied to the bombardment of crystals with fast electrons, i.e., to a primary energy transfer E_A of several tens of eV /26/. The experimental verification of the "thermal disordering" of relatively large regions by particle bombardment of crystals should be based on the facilitation, in that case, of impurity diffusion and self-diffusion.

4. Defect Production by Ionization Processes.

In addition to the displacement of atoms by collisions, which is predominant during irradiation with high energy particles, radiation defects can be produced by X-ray irradiation of ionic dielectric crystals /27, 28/. The production of defects in semiconductor compounds by an ionization process has been confirmed experimentally.

A model of defect production by ionization has been suggested by Varley /28, 29/, who believes that the probability of repeated ionization of negative halide ions in alkali metal halide crystals is rather high. Since newly created positive ions are surrounded by positive ions of the alkali metals, the system is electrostatically unstable. If the multiple ionization lasts sufficiently long, the ion can be displaced from its site into an interstice, for instance, as a result of the combined effect of electrostatic forces and of the thermal motion. According to the Varley model the repulsion forces between ions in the interstices and vacancies can be sufficient to stabilize the defects by a further displacement of the interstitial ion and the vacancy.

Unfortunately the Varley model includes parameters which cannot yet be reliably calculated or experimentally determined. The basic conditions for the production of defects by ionization are a high probability of multiple ionization, a sufficient lifetime of multiply-charged ions and stability of the interstitial position of the displaced ion.

Multiple ionization can be caused by the Auger effect following displacement of an electron from one of the outer shells of an atom, or by independent ionizations, for instance, caused by absorption of two photons. In the first case the number of produced defects is a linear function of the radiation intensity, while in the second case the number of such defects is proportional to the square (or to a higher power) of the radiation intensity. Howard, Vosko and Smoluchowski /30/ who studied the probability of the Auger effect in alkali metal halide crystals have found that interaction with electrons, particularly with those from inner shells, can lead to the absorption of electromagnetic radiation. Such absorption is accompanied by emission of photo-electrons which ionize several shells. The probability of multiple ionization in ionic crystals, e.g. in KCl is sufficient for the Varley process to take place (see /31/).

There is no agreement among investigators with regard to the lifetime of multiply charged ions. According to Howard et al. /32/, the capture time of an electron from the conduction band is much longer than the vibration period of the lattice. Dexter, however, believes /33/ that a positive halide ion can be regarded as two localized holes in a valence bond; the holes are separated in space in the course of about 10^{-15} sec. This approach of Dexter has been criticized by Varley.

The applicability of the Varley model, or of its modification, to semiconductors was discussed at the Conference on Radiation Damage in Semiconductors, held in France in 1964* and was found unacceptable, in its present form, for semiconductors in which the defects are formed by electrons or by low-energy photons. However, the fact of defect formation in the near-threshold range and the great influence of ionization on the rate of defect formation in crystals are well established. An extensive investigation of ionization processes that lead to the production of radiation defects in semiconductor crystals is of great theoretical and practical interest and should certainly attract the attention of many investigators.

---

* Interesting ideas about the ionization mechanism of defect production were advanced by the late S.V. Starodubtsev and by A.E. Kiv /34/.

CHAPTER II

EXPERIMENTAL STUDY OF RADIATION DEFECTS IN SEMICONDUCTORS AND CONTROL OF SEMICONDUCTOR PROPERTIES BY IRRADIATION.

As we mentioned above, the primary processes of displacement of atoms from their crystal sites can be represented by various models. However, modern theories do not allow to predict the stability of the defects produced by irradiation. The energy spectrum of local centers produced by irradiation of semiconductors cannot be calculated, although progress has been made with quasi-phenomenological theories of such centers based on experimentally determined parameters /35/. Most experimental investigations of radiation damage in semiconductors have used techniques which are in part related both to solid state physics and to the production of semiconductors with properties controlled by the nature and the spectra of the energy levels of radiation-induced defects. The problems of solid state physics which can be most conveniently studied in semiconductors include the threshold energy of defect production, and the generation and nature of simple and complex radiation damage.

Important results in the field of the physics of real crystals can be obtained by investigations of recovery in semiconductors, caused by annealing of radiation defects. Investigations of the stability of defects also have practical aspects. Indeed, if defects induce new properties in semiconductors (for instance, regions of infrared photoconductivity), it is important to conserve these properties during the entire service life of the semiconductor device. The detrimental influence of defects, for instance, an increase in the rate of bulk recombination, can, in principle, be avoided by using semiconductor materials with unstable defects or materials with a high threshold energy of defect production.

The number of defects produced by hard radiation and the energy levels of such defects can be studied by measuring the electric conductivity and the Hall effect. The data obtained from investigations of crystals of germanium, silicon and other semiconductors, indicate that the radiation defects produced, even by simple electron bombardment, have a complex spectrum of shallow and deep energy levels in the forbidden band. The fact that lattice defects have several model according to which: (*a*) the damaged region is "removed" from the crystal; (*b*) damage is produced and (*c*) the "removed" part of the crystal is restored to its place.

The structural damage leads to a change in volume and induces stresses not only in the damaged region but also in regions farther away. According to Eshelby /96/ a random distribution of point defects can be expected to produce uniform and isotropic stresses in the crystal. In tetrahedral crystals, vacancies and interstitials can be expected to influence the crystal like isotropic tensile stress centers. In other crystals, too, a random distribution can be expected to produce a large number of isotropic tensile defects in the crystal.

In crystals stretched by existing defects we must take into account both the change in the lattice constants and the change in the number of atoms in their natural sites. An X-ray study of the lattice constants produces mean values while a study of the linear dimensions allows one to determine both the change in the lattice constant and the change in the number of atoms in their natural sites. The appearance of an interstitial or of a vacancy is associated with a change in volume, equivalent to the introduction of an atom in a lattice point on the surface of the specimen or to the removal of such an atom.

If we designate by Ω the atomic volume of a crystal and by $f_i\Omega$ and $f_v\Omega$ — the expansion associated with the appearance of a single interstitial or a single vacancy respectively, the total relative change in the length L and in the lattice constant α can be calculated from:

$$\frac{3\Delta L}{L} = c_i\,(f_i - 1) + c_v(f_v + 1); \qquad (72)$$

$$\frac{3\Delta\alpha}{\alpha} = c_if_i + c_vf_v, \qquad (73)$$

where c_i and c_v — are the relative concentrations of interstitials and vacancies. From the last two equations we have:

$$c_v - c_i = 3\left(\frac{\Delta L}{L} - \frac{\Delta\alpha}{\alpha}\right). \qquad (74)$$

If $c_i \cong 0$ (for instance, in some metals) it is possible to determine the concentration of vacancies, without knowing the values of f_i and f_v, by measuring simultaneously $\Delta L/L$ and $\Delta\alpha/\alpha$, at high temperatures.

According to recent data, the accuracy of determination of $\Delta L/L$ and $\Delta\alpha/\alpha$ is better than 10^{-5} and therefore experiments carried out with sufficiently high irradiation doses and at low temperatures can in principle yield independent data on the number of defects.

Data on isolated (point) defects. As we mentioned above, electrons or photons with an energy of several MeV can produce pairs of interstitials and vacancies. If we disregard annealing (recovery) then $c_i = c_v \equiv c_p$ and from the above equations we have:

$$3\,\frac{\Delta L}{L} = 3\,\frac{\Delta\alpha}{\alpha} = c_p\,(f_i + f_v). \tag{75}$$

Thus, a determination of the length of crystals can be expected to yield data on the sum $f_i + f_v$ provided that we know the concentration of defects. Vook carried out such measurements with crystals of germanium and silicon /97/ and with semiconductor compounds InSb and GaAs /98/.

Irradiation of semiconductors at $50-86^\circ$K with 2 MeV electrons does not change the dimensions of germanium and silicon crystals at doses up to $2 \cdot 10^{19}$ elec/cm^2 while the dimensions (length) of InSb and GaAs specimens increase linearly. For bombardment of InAs and GaAs with 2 MeV electrons the value of $\Delta L/L$ is about 10^{-23}. An estimation of the value of $(f_i + f_v)$ for germanium and silicon, based on the assumption that defects are not annealed, has been made in /94/. However, this assumption is contradicted by the reliable data of MacKay et al /61/, who irradiated germanium with electrons at lower temperatures and still found that annealing takes place.

If the temperature of electron irradiated InSb specimens is increased to between 50 and 170°K this leads to "reverse annealing" accompanied by additional expansion of the specimens. The "reverse annealing" can be due either to rearrangement of defects or to the annealing of defects of a certain type which in the absence of other defects leads to contraction of the specimen.

A theoretical calculation /99/ showed that the value of f_v for semiconductors of group IV of the periodic system should be negative. However, it is possible that in compounds of the $A_{III}B_V$ type vacancies cause an expansion of the crystal in the same way as the repulsion between four neighboring atoms leads to an expansion of ionic crystals /100/.

Pearson *et al.* /101/, measured the lattice constants of germanium and gallium arsenide crystals, quenched from elevated temperatures, using the method of Kossel lines and obtained results which agree qualitatively with data on electron irradiated specimens. The quenched germanium crystals did not change their volume but GaAs expanded due to the influence of vacancies. The activation energy for vacancy production was close to 2 eV.

Radiation-induced defect clusters in germanium and silicon. Vook and Ballufi /102/ who measured the linear expansion of germanium crystals irradiated with 10 MeV deuterons at 25 and 85° K, found that those particles are more effective than electrons. The change in length was proportional to the radiation dose and was almost the same at 25 and 85° K. The authors calculated that the mean number of displaced atoms in defect clusters, due to irradiation with 10 MeV deuterons, was close to six $(E_d = 30\,\mathrm{eV})$. The maximum number of displaced atoms in a cluster exceeded 100. Clusters containing N defects lead approximately to an N-fold tension in the crystal. It is believed /102/ that the qualitative difference between the effect of deuteron irradiation and electron irradiation is due, first of all, to the existence of clusters which contain the largest possible groups of defects.

Pierce who irradiated tunnel diodes (germanium and silicon) with neutrons at 90°K, found that a continuum of energetic levels is produced in germanium immediately after irradiation. This continuum is due, apparently, to the interaction between defects within large clusters /103/. If the specimens are annealed at between 200 and 375°K, this continuum decomposes into discrete levels, which indicates a spreading of defects, from the close clusters, over the crystal. Obviously, the migration of defects should be accompanied by recombination of the components of Frenkel pairs.

A study of neutron irradiated silicon has shown that the clustering of defects here is less pronounced than in germanium. This is due, apparently, to the fact that the paths of displaced atoms and, consequently, the regions in which cascade displacements take place in light substances, are rather large. In addition, in crystals with wide forbidden gaps (1.15 eV for silicon and 0.65 eV for germanium) the wave functions of the electron states of defects are, usually, more localized.

Existing data on the dimensions and lattice constants of germanium crystals, irradiated with neutrons and deuterons /94, 104/, indicate that the defects in germanium after irradiation at room temperature or at higher temperatures consist either of vacancies or more complex forms, i.e., vacancy clusters.

Data on the existence and dimensions of defect clusters can be obtained by small angle X-ray scattering. Such scattering by clusters occurs if the electron densities in the damaged regions differ greatly from the electron densities in the unaffected part of the crystal.

Fujita and Gonzer found strong small-angle X-ray scattering in germanium irradiated with deuterons at 90°K /105, 106/. An analysis of the angular distribution of the intensity of X-ray scattering led the authors to the conclusion that the mean radius of damaged regions is about 35Å. The authors believed that those regions are produced as a result of local melting of the germanium crystal which retains the structure of the molten material. The volume of molten germanium is 9% smaller than that of solid germanium. However, this theory has not been experimentally confirmed. On the contrary, more recent measurements of the dimension of irradiated germanium contradict this theory.

It seems more natural to assume that the X-ray scattering is due to changes in electron density in space charge regions around the defects. A model of such defects was analyzed by Gossick /86/ and Cleland and Crawford /89/.

Radiation-induced defect clusters in semiconductor compounds of the $A_{III}B_V$ type. As we mentioned above, a large fraction of the energy of a primary displaced atom is converted into vibrations of the crystal lattice which are, at the beginning, concentrated within a limited, approximately spherical, region. We still do not know whether local pulse melting takes place in semiconductor compounds irradiated with fast neutrons or heavy particles.

Gonser and Okkelse, who irradiated GaSb and InSb crystals with 12 Mev deuterons at liquid nitrogen temperature, investigated the damage by small angle X-ray reflections /106/. These authors believe that local contractions can take place in the crystals due to fast solidification of the regions, which retain the structure of the liquid phase; however, other studies indicate an expansion of deuteron-irradiated GaSb and InSb specimens /107/.

Local melting of semiconductor crystals can cause decomposition of the compound and precipitation of the components in the form of "colloidal" inclusions. This assumption has not been confirmed experimentally.

Electron microscopy methods. Electron microscopy investigations of radiation defects have been systematically used for the study of germanium. The observations were related only to defects that are stable at room temperature. Parsons *et al.* investigated thin (about 700Å) germanium films which were vapor deposited and heat treated before irradiation. The heat treatment increased the surface area of single crystals /108/. The specimens were examined under an electron microscope before and after neutron irradiation in a reactor. A method was developed to examine the same region of the film before and after irradiation. The photographs taken after irradiation show clearly the damaged regions. The number of those regions is approximately equal to the calculated number of displaced germanium atoms with a high primary energy. The shape of defect clusters could not be determined, because of the insufficient resolving power of the microscope. The mean diameter of the clusters was close to 50Å.

Parsons also studied defects produced in germanium films, bombarded with 100 keV oxygen ions. After irradiation with small doses $(10^9-5\cdot10^{10}$ ion/cm^2) the film had spots, with a damaged structure, 90Å in diameter. After irradiation with high doses these spots were joined into large areas.

Electron diffraction patterns of films before irradiation showed a system of spots typical of crystalline germanium. Electron diffraction patterns of films bombarded with ions showed rings characteristic of amorphous germanium. Patterns of films irradiated with high doses had no spots but only rings.

Bertolotti et al investigated radiation damage regions on the surface of germanium, using chemical etching and the replica method /109/. *N*-type germanium with a resistivity of 1 ohm·cm was irradiated with $5\cdot10^{12}$ cm^{-2} 14 MeV neutrons at room temperature. The etched (111) surfaces were studied before and after irradiation.

Shades were produced by quartz sputtering. The replicas taken from the surface of irradiated but unetched germanium showed no changes. However, after irradiation and etching the replicas showed clearly damaged regions with a diameter from 50 to 100Å (assuming that the regions were of the *p*-type) surrounded by 500—1000 Å pits. Those dimensions agree well with the Gossick model and the number of defects agrees satisfactorily with theoretical calculations.

Very interesting were experiments that confirmed the ordering, i.e., crystallization, of amorphous germanium in displacement spikes. Thin films of metastable amorphous germanium (100—700 Å) were deposited by sputtering on table salt and bombarded with neutrons and with accelerated xenon ions at room temperature. A comparison between electron microscope photographs made before and after irradiation, showed that the irradiated films contained crystalline grains.

The mean diameter of grains, produced by neutron bombardment, is close to 200 Å. The authors believe that the increase in the kinetic energy of atoms in the region of strong excitations (displacement spike) leads to recrystallization of the region in which a displacement cascade develops.

8. Influence of Radiation Defects on the Thermal Conductivity of Semiconductors.

Berman /110/ who studied neutron irradiated quartz, found that the thermal conductivity is greatly influenced by the concentration of defects in the crystal. Detailed investigations of the influence of radiation-induced defects on the thermal conductivity have been carried out on dielectric and ionic crystals /5/.

Recently those methods were applied to semiconductors. Data on the temperature dependence of the thermal conductivity can be used to obtain information on the microstructure of defects. In many semiconductors primary defects migrate in the crystal, interact with impurities and are annealed even at low temperatures. Therefore, the studies of primary defects, i.e. irradiation of specimens and thermal conductivity measurements, had to be carried out at low temperatures.

Such measurements were carried out by Vook (on GaAs and on InSb /111/) who found that the excess thermal resistivity of GaAs crystals (dimensions $cm.deg.w^{-1}$) increases linearly with increasing 2 MeV electron flux at 70°K. This increase, accompanied by a linear expansion of GaAs specimens, agrees with the theoretical analysis of photon scattering in stressed crystals /112/. About 70% of the excess thermal resistance disappears upon annealing at up to 325°K. This fact agrees well with the observations of Pearson et al, according to which in quenched GaAs specimens about 65% of the produced vacancies are annealed within 24 hours at room temperature. Vook believes that the temperature dependence of the thermal conductivity indicates that only point defects are annealed below 325°K while association or precipitation of defects occurs at 325–575°K. Those results do not confirm the assumption of Aukerman /113/ that annealing of irradiated GaAs at 500°K is associated with the annihilation of Frenkel pairs.

In InSb the excess thermal resistance increases slower than the electron flux.

Electron irradiation decreases the thermal conductivity of InSb much less than that of GaAs /94/. A theoretical explanation for this phenomenon is difficult.

The great sensitivity of thermal conductivity measurements showed that defects are produced in InSb by low-energy electrons or X-rays, even though the atoms cannot be displaced by the impact of scattered electrons /114/. This fact indicates that at small energies the displacement is influenced not only by collisions but also by processes associated, among others, with ionization, i.e., with changes in the charged state, and with the stability of the atom in its site.

Data on the changes in the thermal conductivity of germanium and silicon after neutron irradiation in a reactor at low temperatures and at room temperature can be found in /115, 116/. An analysis of the temperature dependence of thermal conductivity indicates that both point defects and clusters are formed.

An analysis of the changes in linear dimensions (length) of irradiated semiconductors, and X-ray diffraction studies, were used to differentiate between the production of isolated point defects and defect aggregates or regions of local melting. In germanium and silicon, stresses associated with point defects are relatively small, while in semiconductors with partially ionic bonds, for instance, in $A_{III}B_V$ compounds, the stresses due to the interactions of ions are considerable.

Stresses arising in radiation-induced clusters in germanium are very complex and lead to a relative intensification of stresses in the whole crystal. In silicon the stresses are smaller, which agrees with data obtained by photoconductivity measurements.

Electron microscopy methods and thermal conductivity measurement are useful auxiliary means for the determination of the structure and stability of radiation defects.

9. Radiation Doping of Semiconductors.

Radiation doping of semiconductors can be used to control their properties. Today extensive investigations are carried out on the doping of germanium and silicon by ion implantation. Theories are developed and experiments are carried out on the penetration of charged particles into a crystal due to channeling in certain crystal directions. It has been found that irradiation can considerably accelerate the diffusion of impurities in crystals. Irradiation (particularly with neutrons) of semiconductors frequently produces impurity atoms as a result of nuclear reactions (transmutation doping). Those problems would be discussed in brief.

Research in those fields expands constantly; however, most of the research, particularly on ion implantation, is of an empirical nature. In this chapter we shall discuss only the aspects that are most important from the practical and theoretical points of view.

Ion implantation of dopants. The control of semiconductor properties through implantation of dopants by fast-ion bombardment was studied by M.M. Bredov, N.M. Okuneva, V.L. Lepilin and A.V. Nuromskii in 1957–1961 /117/. They made several theoretical assumptions and carried out experiments which showed that donor atoms (Li) implanted into a p-type semiconductor (Si) can change the nature of conductivity. Earlier, P. Ohl /118/ used ion bombardment for "etching". The operating conditions and the type of ions (usually He^+, N^+, Ar^+, etc.), were chosen empirically. Ion etching for surface cleaning, although of great practical interest, is beyond the scope of this book.

The mean paths of certain important dopant ions (P, B) can be approximately calculated using the ordinary equations of the theory of ionization losses /119/. For doping of silicon semiconductors (in order to produce p-n junctions in the vicinity of the crystal surface) such calculations are sufficient.

An important example of the practical use of ion implantation is the production of photoelectric converters (elements of solar batteries). V.M. Gusev *et al.* /120/ used for this purpose an ion beam apparatus, with an acceleration voltage of 30 keV, a magnetic separator, and an arc-discharge source.

An approximate assessment has shown that in phosphorus-implanted silicon the depth of the p-n junction is equal to several tenths of one micron. An analysis of the spectral characteristics of silicon photoelectric cells has shown, as expected, that the properties of the crystal beneath the *p-n* junction are almost not influenced by the ion implantation. This means that by this method (unlike other methods associated with a high temperature treatment) it is possible to produce a non-equilibrium carrier with a long lifetime. One of the unavoidable problems of ion implantation is the production of radiation-induced defects, part of which are stabilized and affect the properties of the semiconductor.

Implantation of phosphorus ions in silicon at room temperature and even up to $500°C$, results in defects with deep carrier capture levels in the forbidden gap, which increase the resistivity of the crystal. Apparently, because of screening of part of the implanted atoms in high-resistance regions, only a fraction of those atoms remains electrically active in non-annealed silicon. Annealing at $500-600°C$ considerably reduces the resistivity of silicon doped with large doses and makes it possible to produce *p-n* junctions with smaller doses. This is due to annealing of radiation defects and to transitions of implanted but unactive phosphorus atoms into the electrically active (donor) state /121/.

Reliable data based on diagonal cross-sections have shown that after annealing the *p-n* junctions were at a depth of $0.75-1.1\ \mu$, i.e., $15-20$ times deeper than the average depth of penetration of 30 keV phosphorus ions. It is believed that heating of ion-implanted specimens causes annealing and diffusion of implanted atoms. The high diffusion velocity indicates that the displacement of impurity atoms is accelerated by the presence of vacancies produced by the ion beam. Similar effects were extensively studied in France by Baruch and Pfister /122/.

Ferber investigated the implantation of H^+, N^+, O^{2+}, Ne^+ and Xe^+ ions in silicon, and found a resistivity increase which he attributed to the formation of deep donor states /123/. According to the theories of radiation damage, clusters of point defects would appear at the end of the ion path and their total number, at least in the initial moment, greatly exceeds the number of implanted ions. P.V. Pavlov *et al.* /124/ found that above a certain critical dose of ion irradiation a damaged layer is produced on the surface of silicon and germanium. The thickness of that layer increases with increasing dose. The authors believe that the layer is amorphous. Similar layers have been found by other investigators /125/.

At the International Conference on Radiation Defects in Semiconductors /126/, P.V. Pavlov *et al.* reported interesting data on the distribution of electrically active atoms of boron and silicon as a function of the orientation of the ion beam relative to the crystallographic axis. Experimentally confirmed calculations have shown that the "centers of gravity" and the maximum concentrations of defects are closer to the surface than those of the dopant atoms introduced by ion implantation. The activation energy of implanted boron atoms is, even after annealing, higher than that of boron introduced during the growth of single crystals. No such differences have been found for phosphorus.

Calculated and experimental data on the distribution of boron ions implanted in single-crystal silicon by bombardment along different crystallographic axes are given in Figure 33.

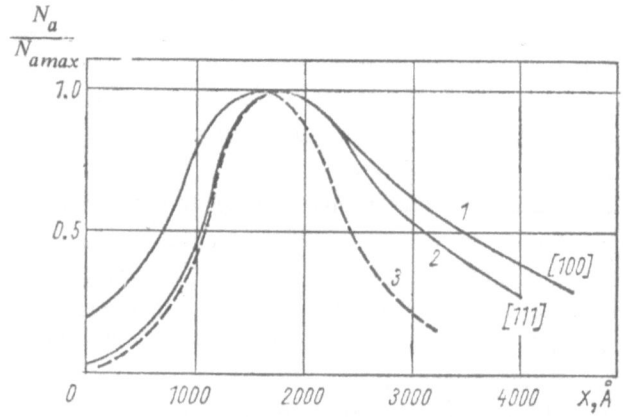

Figure 33. Experimentally determined distribution of boron atoms in silicon (x–distance from the surface). 40 keV ions; the ion beam was directed along the [100] (1) or [111] (2) axis (3-calculated curve) /126/.

Ion implantation is a very useful method for introduction of dopants, e.g., in research. The production of semiconductor diamonds necessitates the introduction of impurities at elevated temperatures and under a high pressure. Using ion implantation Soviet workers /127/ have produced natural diamonds with *n*-type conductivity (Li-doped) or *p*-type conductivity (B-doped). In conclusion we would quote typical data from a modern ion beam apparatus used for dopant implantation in crystals at the Nuclear Research Center, Harwell (England) /128/. In 1967 the maximum ion energy was increased to

400 keV, Results obtained with a beam current of 10 μA to 1 mA are shown in Table 3 (for the implantation of P, B, N, Cl, Ta, W).

Table 3. Operating conditions of a modern ion beam apparatus /127/

Operating conditions	Normal process			Postdeflection acceleration on the collector		
	Ion charge			Ion charge		
	1+	2+	3+	1+	2+	3+
Maximum energy keV	45	90	135	145	290	435
Current μA	1000	50	1	100	50	1
Mean ion flux 10^4 cm$^{-2}\cdot$sec^{-1}	60	3	0.06	6	3	0.06
Rate of doping ion$\cdot$cm$^{-3}\cdot$sec^{-1} (at a thickness of 1μ)	$6\cdot10^{19}$	$3\cdot10^{18}$	$6\cdot10^{16}$	$6\cdot10^{18}$	$3\cdot10^{18}$	$6\cdot10^{16}$
Surface area of the beam on the article	Rectangle (4x4 cm)			Circle, 2 cm in diameter, limited by the aperture of the electrostatic lens		
Angular resolution	$\pm1°$ for ordinary implantation of ions in small crystals (about 1 cm^2) the beam must be collimated with an accuracy of up to 0.1°, as for instance in experiments on channeling					

The ion implantation method holds great promise for the solution of certain problems in semiconductor electronics, particularly for the production of integrated circuits. In cases where it is necessary to produce strong stable joints between two materials (for instance in mirror coatings), ion implantation is superior to all other methods.

We have discussed above irradiation which is not influenced by the mutual orientation of the crystal axis and the direction of ion penetration into semiconductors.

Channeling, which we shall discuss below, was discovered in 1963.

"Channeling" of charged particles in crystals. A theoretical analysis of the motion of heavy ions with an energy of several keV, has shown that unlike strongly scattered electrons, and because of the periodical

structure of the crystal, such ions can be focused in a way resembling "hard" focusing in certain accelerators. This phenomenon is called channeling; it shows a correlation between the free path and the ionization losses on one hand and the mutual orientation of the direction of the accelerated ions in the crystal and the crystal axis on the other* /129-131/.

If a particle enters a channel (confined between chains of regularly placed atoms) at a small angle with respect to the axis of the channel (Figure 34) it will move forward with periodic oscillations in the

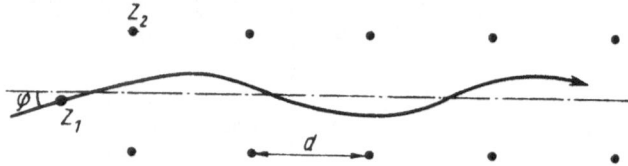

Figure 34. Motion of charged particles along natural "channels" in crystal lattices.

vicinity of the axis /132/. Focusing is a result of interactions between the electric field of the moving atom and of atoms of the crystal lattice. The De Broglie wavelength of heavy ions is smaller than their lattice constant "a" and therefore classical theories can be used for the analysis of the basic features of channeling. For ideal rigid lattices, consisting of identical atoms with a nuclear charge $Z_2 q$, the maximum angle φ_{max} at which a particle can move along a channel can be found from:

$$\varphi_{max} \simeq 1.5 \sqrt{\frac{b}{d}},$$

where $b = q^2 Z_1 Z_2 / E$, and $Z_1 q$ is the charge of the moving particle; E is its kinetic energy /133/ and d the distance between atoms in the chain.

A beam passing through a crystal splits into two currents: a current of particles which enter the crystal at an angle $\varphi > \varphi_{max}$ and undergo the usual multiple scattering, and a current of particles with an angle of incidence (relative to the channel axis) $\varphi < \varphi_{max}$, which continues to move without coming closer to the chains. The scattering of such

---

* Stark, who in 1912 predicted such an effect, believed that it would be the free electrons that would move through the crystal.

particles is greatly influenced by the thermal vibrations of the lattices and by their imperfections. Obviously, a few particles from the current passing strictly along the channel, which are scattered by very close collisions with crystal atoms, would leave the beam. If we designate the Thomas-Fermi scattering radius as $a' = 0.885a_0Z_2^{-1/2}$, where a_0 is the radius of the first orbital of a hydrogen atom, then the ideal number of scattered particles in a unit volume will be equal to: $X_{min} \cong \pi a'^2 Nd$, where N is the number of atoms per unit volume*.

As we mentioned above, the energy losses of channeled particles are smaller than the mean ionization energy losses of randomly moving particles.

The channeled particles have abnormally long paths, since the distribution density of electrons in the vicinity of the channel axis is much smaller than average and therefore the interaction of the field of the moving particles with those electrons is weak.

In addition to this effect, which is most important from the standpoint of ion implantation, the abnormal travel of particles has other effects /134/. Thus, the particle trajectory is greatly influenced by the yield of nuclear reactions, which can take place as a result of the interaction of particles (e.g. protons) with nuclei in the crystal /135/.

Channeling of protons in silicon single crystals was studied by Dearnaley /136/ and Eisen /137/.

The energy of protons in the Eisen experiments varied from 60 to 400 keV. About 0.001% of the 60 keV protons impinging on specimens 2μ thick passed through the crystal. 400 keV protons also passed through the crystal even without channeling, i.e. by random motion. The proton current, at the optimum orientation of the beam along the [100] axis, was equal to 0.6 of the total current of incident protons. The author investigated also the dependence of the intensity of the current passing through the crystal on its orientation relative to the direction of proton incidence. For 400 keV protons the width of the channel, corresponding to half the maximum intensity, was equal to 1.6°. The angular width increased almost proportionally to the reciprocal of the square root of proton energy. This agrees with the Lindhard theory /133/.

Channeling can influence the distribution of radiation defects in crystals, since the probability of scattering of channeled particles at large angles and, consequently, the production of defects, is much

---

* This assessment is true if $Z_1 \ll Z_2$

smaller than that for randomly moving particles.

In the previous chapter we mentioned the formation of amorphous layers on silicon and germanium by ion beam bombardment. Sometimes such amorphous layers are visible because of the change in the coefficient of optical reflection /138/. The amorphous nature of the visible layers has been confirmed by the fact that they disappear at the same temperature (670°C) at which a normal electron diffraction pattern is obtained. The formation of a visible layer is the result of accumulation of a large number of radiation defects. No such amorphous layer is formed if silicon is bombarded by a collimated beam of 40 keV neon ions at angles of incidence corresponding to ion channeling. /138/.

Very important practical observations have been made by Moak *et al.* /139/ who used silicon surface-barrier detectors for the registration of pulses from high-energy bromine and iodine ions. Randomly moving ions lose some of their energy on nuclear collisions which do not lead to ionization. However, if the ions move along crystal axes with small lattice constants the ionization pulses are most intense since no energy is lost on scattering /125/. It is also possible that because of the long path and the smaller losses of ion energy (dE/dx), recombination losses in the plasma within the ionization column during channeling are also very small.

Semiconductor detectors used for recording of heavy ions have a short service life because of the accumulation of defects. However, this lifetime can be greatly prolonged by channeling.

A study of single crystals of tungsten /140/ and silicon /141/, has shown that a small fraction of impurity atoms penetrate very deeply into the crystals. However, the depth of penetration is not influenced by the mutual orientation of the ion beam and of the crystal axis. It has been found /142/ that this penetration is due to stimulation of impurity centers (see also /122/). According to /142/ a large part of the implanted atoms are trapped at interstices after which they diffuse along interstitial sites until they encounter vacancies and occupy regular atom sites.

Doping by nuclear reactions (transmutation doping). Chemical impurities can be produced in crystals by nuclear reactions. Particularly important, from both the theoretical and practical standpoints, is the interaction with neutrons which produces a bulk effect*. Studies of great theoretical importance for the elucidation of the nature of local electron states of impurity centers /72/ became possible after very pure single crystals of germanium and silicon became available and after reliable data were obtained on the cross-sections of nuclear reactions for certain isotopes (for instance, germanium). In order to separate the residual effects, associated with the chemical impurities, germanium single crystals were irradiated with slow neutrons and, after removing from the reactor, were given a long (24 hour) anneal at $450°$ C. No further changes in electric conductivity were found after this process. The annealing led in the first place to a recovery of radiation defects which had been induced by a background of γ-irradiation.

The most important types of nuclear reactions involving germanium isotopes are given in Table 4 /72/. The capture cross sections of slow neutrons in germanium isotopes are given in Table 5. As can be seen, nuclear transformations taking place in germanium bombarded with slow neutrons lead to the formation of atoms of gallium (acceptors) and of arsenic (donors). The concentration of those elements can be calculated using a simple equation:

$$N_i = \Phi \sigma N_{Ge} P_t,$$

where Φ is the integrated neutron flux; σ is the capture cross section; N_{Ge} is the number of germanium atoms per cm^3; P_t is the content of the isotope.

Calculated data on the number of new impurity centers agree well with data on the Hall effect and on the electrical conductivity.

The same method was used by N.P. Kekelidze /143/ for the determination of the concentration of donor and acceptor centers in germanium, on the basis of data on the Hall effect at low temperatures. Nuclear doping must be taken into account in semiconductor devices operating under conditions of strong radiation. Attempts have been made to dope silicon with a phosphorus isotope /144/ in order to compensate the residual concentration of boron (acceptor) atoms produced by neutron bombardment in silicon detectors with a large active volume.

Unfortunately the non-uniform distribution of residual boron atoms led to an increase in the relative heterogeneity at high fluxes and the

* In semiconductors containing components with a large capture cross section (Cd) the impurities and defects become nonuniformily distributed also by neutron irradiation.

Table 4. Nuclear reactions in germanium.

Nuclear reaction

A. Reactions with deuterons and slow neutrons

Isotope	Content, %	(d,p)	(n,γ)	(d,n)
Ge^{70}	21.2	$Ge^{71} \xrightarrow{11.4\ Days} Ga^{71}$	$As^{71} \xrightarrow{50\ Hours} Ge^{71}$	$\xrightarrow{11.4\ Days} Ga^{71}$
Ge^{72}	27.3	Ge^{73}	$As^{73} \xrightarrow{76\ Days} Ge^{73}$	
Ge^{73}	7.9	Ge^{74}	$As^{74} \xrightarrow{17.5\ Days} \left\{\begin{array}{l} Ge^{74} \\ Se^{74} \end{array}\right.$	
Ge^{74}	37.1	$Ge^{75} \xrightarrow{82\ Minutes} As^{75}$	As^{75}	
Ge^{76}	6.5	$Ge^{77} \xrightarrow{12\ Hours\ 50\ Seconds}$	$As^{77} \xrightarrow{40\ Hours} Se^{77}$	$\xrightarrow{40\ Hours} Se^{77}$

B.: Reactions with fast neutrons

Isotope	Content, %	(n,p)	$(n,2n)$	(n,α)
Ge^{70}	21.2	$Ga^{70} \xrightarrow{20\ Minutes} Ge^{70}$	$Ge^{69} \xrightarrow{40\ Hours} Ga^{69}$	Zn^{67}
Ge^{72}	27.3	$Ga^{72} \xrightarrow{14\ Hours} Ge^{72}$	$Ge^{71} \xrightarrow{11.4\ Days} Ga^{71}$	$Zn^{69} \xrightarrow{15\ Hours} Ga^{69}$
Ge^{73}	7.9	$Ga^{73} \xrightarrow{5\ Hours} Ge^{73}$	Ge^{72}	Zn^{70}
Ge^{74}	37.1	$Ga^{74} \longrightarrow Ge^{74}$	Ge^{73}	$Zn^{71} \xrightarrow{3\ Minutes} Ga^{71}$
Ge^{76}	6.5	$Ga^{76} \longrightarrow Ge^{76}$	$Ge^{75} \xrightarrow{82\ Minutes} As^{75}$	$Zn^{73} \xrightarrow{2\ Minutes} Ga^{73} \xrightarrow{5\ Hours} Ge^{73}$

Table 5. Capture cross sections of germanium isotopes for slow neutrons.

Isotope	Content %	Capture cross section, barn		Stable element
		for isotope	for atom	
Ge[70]	21.2	3.25	0.69	Ga
Ge[72]	27.3	0.94	0.26	Ge
Ge[73]	7.9	13.69	1.08	Ge
Ge[74]	37.1	0.60	0.22	As
Ge[76]	6.5	0.35	0.02	Se

method of nuclear doping appeared therefore to be less perfect than the method of drifting of donor impurity ions (Li) in an electric field, which produces uniform electrical properties in silicon /10/.

The recoil of nuclei capturing neutrons, which is accompanied by the emission of γ-quanta can produce a considerable number of radiation defects /145, 146/. According to /146/ the number of defects, produced by nuclear recoil, during irradiation of silicon in reactors is comparable with the number of defects produced by scattering of fast neutrons. The energy of nuclear recoil can be calculated from:

$$E_A = \frac{537}{A}(h\nu)^2 \text{ eV},$$

where $h\nu$ is the energy of γ-quanta emitted by nuclei, MeV; A is the atomic weight of the nucleus. The number of defects can be calculated only on the basis of complete data about the system of energy transitions of excited nuclei, which can be found, among others, in /147/.

Indium antimonide is widely used as a material which can be doped by irradiation with slow neutrons. Indium has a large effective activation cross section ·by slow neutrons (140 barn). The In[115] isotope, which comprises 95.7% of the natural mixture, can be converted by a nuclear reaction (n, γ) into a stable Sn[116] isotope. In the natural sites of the crystal lattice those atoms are donor centers.

The (n, γ) reaction converts antimony into tellurium but the effective cross section of this reaction is very small (0.013 barn) and therefore the conversion of indium is much greater. L.K. Vodopiyanov and N.I. Kurdiani /148/ who investigated the influence of nuclear doping on the optical and electrical properties of InSb have elucidated the filling of the conduction band by carriers which shift the edge of the main optical absorption band towards shorter wavelengths (Burstein effect). Irradiation of acceptor-doped InSb was used to study the theoretically predicted /149, 150/ changes in the band structure of highly doped semiconductors by nuclear implantation of Si (donor) and by accurate compensation of the concentration of initial acceptors.

CHAPTER III

IONIZATION IN SEMICONDUCTORS AS A RESULT OF THE STOPPING OF CHARGED PARTICLES, ABSORPTION AND SCATTERING OF PHOTONS

The production of equilibrium carriers (electrons and holes) in semiconductors and in dielectrics by charged particles and by high energy γ-rays is of great importance because of the conversion of the energy of those particles into light and electrical energy and because of the possiblity of using semiconductors for detection and determination of the type of radiation, and of the energy of charged particles (or of γ -rays) in dosimetry.

Let us now consider the physical phenomena which convert "hard" radiation into energy (of non-equilibrium carriers) or into thermal excitation energy, i.e. into vibration of atoms in a crystal lattice.

1. Energy Losses on Ionization

Energy losses by charged particles. Any charged particles: electrons, positrons, α-particles, protons, mesons, etc. which move through a substance gradually lose their kinetic energy; this process of energy loss is called particle stopping. The main energy losses are due to interaction between the electric field of the charged particles with electrons of atoms in the substance. The excitation energy of an electron can be of any magnitude if the electron is outside the atom or it can acquire only certain discreet values if the electron is not free but only passes into an excited state within the atom.

Since the stopping of charged particles in gases leads to ionization, i.e. to the production of free electrons and of positive ions, the energy losses are called ionization losses. Stopping of fast charged articles in dense media, i.e. in liquids or in solids, is also due chiefly to energy transfer to electrons in the atom. The term "internal ionization" is associated not only with stopping of particles during ionization in gases but also in solids and in liquids. In semiconductors or in dielectrics, which can also be considered as semiconductors with a wide forbidden band, internal ionization corresponds to the transition of a valence electron into the conduction band*. Thus, excess carriers and holes are "free" only within the crystal, except for a small number of electrons, which are freed in the vicinity of the surface and can overcome the potential barrier and leave the crystal.

The problem of ionization losses from charged particles is one of the fundamental problems of nuclear physics. The dependence of the specific energy losses E, which are determined by dE/dx (x — coordinate of particle on its trajectory), on the charge and on the velocity of particles and on the parameters of the stopping material has been studied in detail /9,151/. A summary of experimental studies of the stopping of particles can be found in /9/. A review of the subject can also be found in the book of Dearnaley *et al.*/10/. We shall discuss briefly the basic theories of energy losses from charged particles, X-rays and γ-radiation, and concern ourselves with the results of energy transfer to electrons in semiconductors. The main purpose of the above studies was to understand the behavior of particles. Little attention was given, until recently, to the effects of the passage of those particles on matter.

In order to present a valid model of the interaction between a charged particle and electrons in an atom we shall use the classical Bohr calculations without the use of quantum mechanics.

Let us assume that an electron of mass m is at a distance b from the trajectory of a particle of mass M, a charge Zq and a velocity v. Let us further assume that the electron is free and during its interaction with the particles it moves so slowly that in a calculation of the electric field of the charge of the particle, which acts on the electron, we can disregard its displacement. In order to calculate the energy, acquired by an electron, we must determine the momentum which is imparted by this particle. The direction of the electrostatic force changes with time. If the mass of the particle is much larger than that of electrons, the direction of its motion and its kinetic energy are little influenced by a single interaction with an electron. Because of symmetry, the component of electron momentum in the direction of particle trajectory is equal to zero, since for every position of the particle to the

* The ionization of impurity centers, which greatly influences the equilibrium electric conductivity of semiconductors, the intrinsic photoelectric effect and the electric breakdown, has little influence on the ionization losses of fast charged particles.

left of point A (Figure 35) in which the momentum is changed in one direction there is a symmetric particle to the right of A, this causes an identical change in the component of the momentum but in an opposite direction, parallel to the trajectory of the particle.

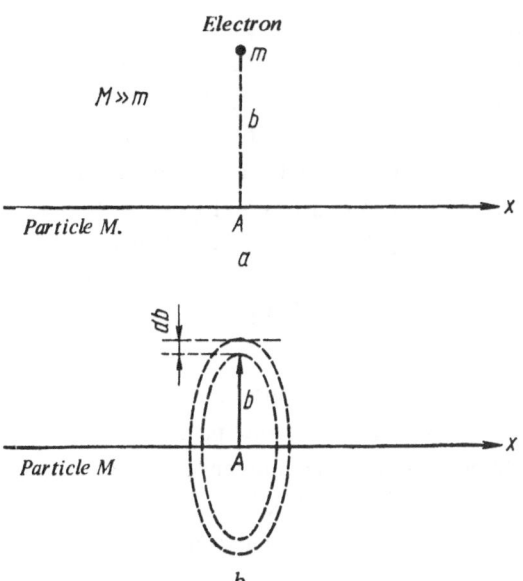

Figure 35. Energy imparted to an electron by a moving particle (classical approximation).

The component of the electron momentum perpendicular to the trajectory of the particle can be calculated from:

$$P_\perp = \int F_\perp \, dt. \qquad (76)$$

The order of magnitude of $P_\perp$ is determined by the product of the time of interaction and of the electrostatic force, i.e.,

$$P_\perp \simeq \frac{Zq^2}{b^2} \cdot \frac{b}{v}. \qquad (77)$$

A more accurate calculation has shown that the momentum, $P = P_\perp$ acquired by the electron, is twice as large, i.e., $P = 2Zq^2/bv$ and, consequently, the kinetic energy ΔE is equal to:

$$\Delta E = \frac{P^2}{2m} = 2\frac{Z^2q^4}{mv^2b^2}. \tag{78}$$

On its way the particle interacts with a large number of electrons with different values of b and, consequently, transfers different fractions of its energy ΔE. A cylindrical layer of unit length, limited by the values of b and $(b+db)$ contains $2\pi Nbdb$ electrons where N is the total number of electrons per cm^3 of substance. The energy lost by particles on the excitation of electrons per unit path can be calculated by summing up the losses in all layers (Figure 35.b) lying between the extreme values of b_{max} and b_{min}, as we shall discuss later. Thus,

$$-\frac{dE}{dx} = \frac{4\pi Z^2q^4N}{mv^2}\ln\frac{b\ max}{b\ min}\ \text{erg.cm}^{-1} \tag{79}$$

If the particle velocity v approaches the velocity of light in vacuum c, the electric field of the particle is "shortened" in the direction of the motion and the component of the field, perpendicular to the trajectory x, increases $1/\sqrt{1-\beta^2}$ times, where $\beta = v/c$. The duration of action of the force on the electron (τ) can be calculated from:

$$\tau = \frac{b}{v}\sqrt{1-\beta^2} \tag{80}$$

Thus, the total momentum and the energy imparted to the electron do not change. The time of interaction between a particle and each of the electrons is $\tau = b/v$. If the value of $1/\tau$ is much smaller than the frequency of vibration of bound electrons v, i.e., if the probability of excitation is small, the energy losses will also be negligibly small.

The maximum radius of interaction b_{max} is determined from the last condition as follows:

$$b_{max} = \frac{v}{\overline{v}\sqrt{1-\beta^2}}, \tag{81}$$

where v is the mean frequency of electron vibrations in atoms. The value of b_{min}, is, according to the classical model, determined by assuming that the velocity imparted to an electron by a heavy particle during a

106

head-on collision cannot exceed $2v$ and therefore the maximum energy of the electron would be $\Delta E_{max} = \frac{1}{2}m(2v)^2$

Thus,

$$b_{min\ cl.} \simeq \frac{Zq^2}{mv^2} .$$

(82)

The theory of quantum mechanics /152/ yields a different value of $b_{min.q}$, which can be written:

$$b_{min.q} \simeq \frac{\hbar\sqrt{1-\beta^2}}{mv} .$$

(83)

Hence, if we use the value of the mean vibration frequency of bound electrons $\bar{v}$, we obtain

$$-\frac{dE}{dx} = \frac{4\pi Z^2 q^4 N}{mv^2} \ln \frac{mv^2}{\hbar\bar{v}(1-\beta^2)} \text{ erg.cm}^{-1}$$

(84)

Usually the value of $\bar{v}$ is replaced by the mean ionization potential of atoms (of the absorbant) $\bar{I}$; a more accurate calculation of $-dE/dx$ using I and other of the above parameters leads to:

$$-\frac{dE}{dx} = \frac{4\pi Z^2 q^4 N}{mv^2} \left(\ln \frac{2mv^2}{\bar{I}(1-\beta^2)} - \beta^2 \right) \text{ erg.cm}^{-1}$$

(85)

which is valid for heavy charged particles.

The free paths of protons and of alpha particles, of different starting energies, in germanium and in silicon, are given in Figure 36/10/. The calculation of the free path of particles using the equation

$$R = -\int_0^{E_0} \frac{dE}{dE/dx} ,$$

where E_0 is the initial energy, is made difficult by the capture of electrons at the end of the trajectory. Therefore, for small energies it is preferable to use experimental data.

The stopping of fast electrons in a substance leads not only to ionization losses of energy but also to electron scattering. Such scattering is due chiefly to the elastic interaction of electrons with the

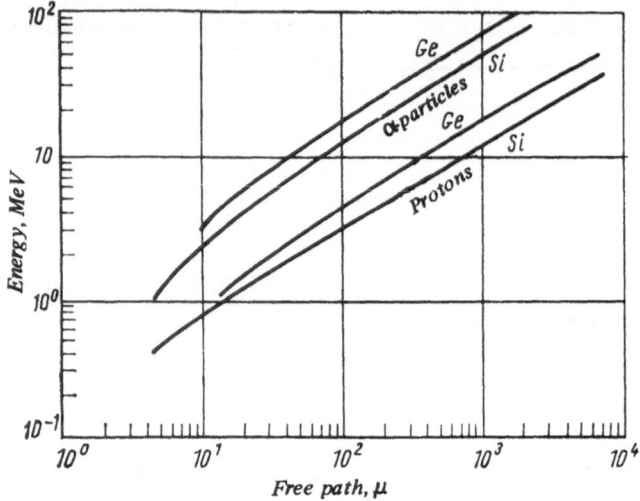

Figure 36. Path lengths of protons and α-particles in germanium and silicon, calculated by the authors of /10/ according to data from /153/.

atomic nuclei, although electrons are also scattered by collisions with bound electrons. Because of the small mass of electrons, they frequently are deflected at large angles. The smaller the energy of an electron, the larger is its mean angular deflection from the initial direction. Because of multiple scattering, the trajectory of electrons has a complex shape and its real length is 1.5—4 times the thickness of the layer traversed by the electron in the initial direction. In calculating the specific ionization we must take into account the fact that it is impossible to distinguish between different electrons after interaction (exchange effects).

For fast electrons Moller proposed the following equation /154/:

$$-\frac{dE}{dx} = \frac{2\pi q^4 N}{mv^2} \left[\ln \frac{mv^2 T}{2\bar{I}^2 (1-\beta^2)} - \ln 2 \left(2\sqrt{1-\beta^2} - 1 + \beta^2 \right) + 1 - \beta^2 \right],$$

(86)

where T is the relativistic kinetic energy of electrons.

Figure 37, a shows the dependence of the specific energy loss from an electron $-dE/dx$ along its trajectory in silicon on the initial energy. The approximate mean ionization potential I is $11.5Z \cdot 1.6 \cdot 10^{12}$ erg for light substances ($Z < 15$) and about $9\underline{Z} \cdot 1.6 \cdot 10^{12}$ erg for heavier substances. Figure 37.*b* shows the dependence of the electron path in silicon on the electron energy.

In addition to energy losses on ionization, fast charged particles also lose energy through bremsstrahlung. This phenomenon is most important for fast electrons. At energies exceeding a critical value $E_{cr.}$, bremsstrahlung losses accompanying the stopping of electrons in the electric field of atomic nuclei are greater than the ionization losses. According to Bethe and Heitler/155/ the ratio A of radiation losses to ionization losses for fast electrons with an energy E_0 is equal to:

$$A = \frac{\left(\dfrac{dE}{dx}\right)_{brem}}{\left(\dfrac{dE}{dx}\right)_{ion}} \simeq \frac{E_0 Z}{1600 mc^2}.$$

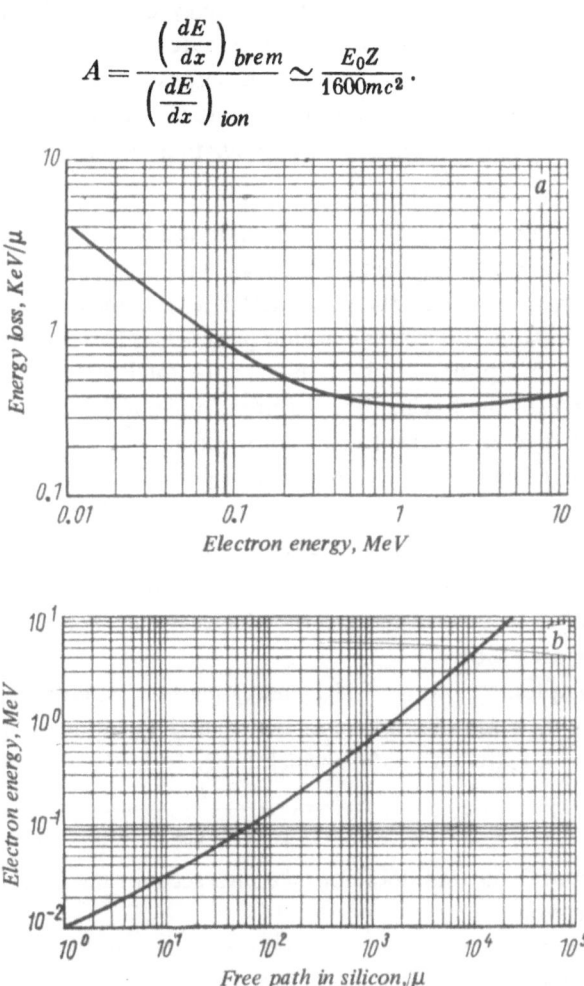

Figure 37. Energy losses from 0.01-10 MeV electrons in silicon (*a*) and free paths of electrons in silicon (*b*).

109

The critical energies, at which losses through ionization and bremsstrahlung are equal, are about 25 MeV for germanium ($Z=32$) and about 57 MeV for silicon ($Z=14$).

In addition to ionization and bremsstrahlung (which cause energy losses) there are other processes in which the energy of fast particles is transferred to the substance. Thus, direct excitation of lattice vibrations is possible in crystals. Seitz/156/ found that such losses are negligibly small compared with ionization losses.

Attempts have been made to elucidate the role of "plasma oscillations" in the transfer of the energy of charged particles to the crystal. However, there is still no definite proof that a relation exists between the so-called characteristic losses of electrons/157/, due to plasma oscillations, and the mean ionization energy, ε i.e. the ratio of energy loss from a particle $\int_{x_1}^{x_2} (dE/dx)dx$ to the number of produced pairs of non-equilibrium electrons and holes.

Ionization as a result of absorption and scattering of γ-radiation and X-rays. The interaction between γ-rays and matter is limited to three basic processes, the photoelectric effect, the Compton effect and the formation of electron-position pairs. The photoelectric effect leads to true absorption; which means that the total energy of the photon is used up for the freeing of an electron from one of the atom shells. The kinetic energy of the freed photoelectron is $E_e = h\nu - E_b$ (E_b is the binding energy of the electron in an atom, which is used for ionization and excitation). The absorption coefficient τ of photons with an energy exceeding the binding energy of electrons in the K-shell is equal to:

$$\tau \simeq NZ^5 (h\nu)^{-3.5} \cdot 10^{-33} \text{ cm}^{-1},$$

where Z is the charge of the nucleus and N is the number of atoms/cm^3/158/.

Compton scattering of high-energy photons on electrons imparts an energy $E_\beta = h\nu - h\nu'$ where $h\nu$ is the initial quantum energy; $h\nu'$ is the energy of scattered photons which depends on the angle of scattering:

$$h\nu' = \frac{h\nu}{1 + (1 - \cos\theta)\dfrac{h\nu}{mc^2}}$$

The above equation does not take into account the binding energy of the electron.

Compton electrons can have any kinetic energy up to the maximum value of $E_{\beta\,max}$, which corresponds to a scattering angle $\theta = 180°$:

$$E_{\beta\,max} = \frac{2(h\nu)^2}{mc^2 + 2h\nu}.$$

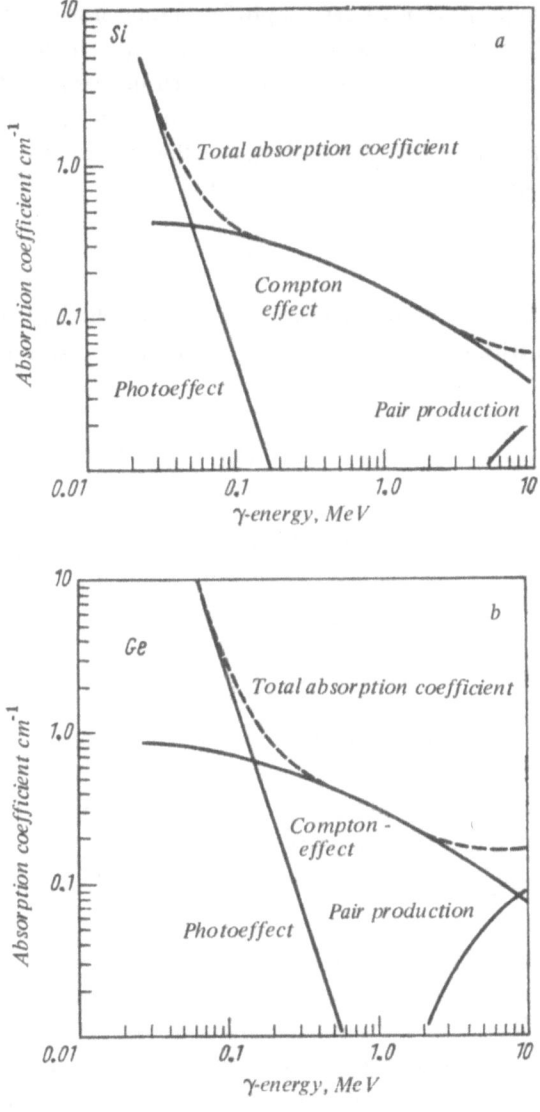

Figure 38. Absorption of γ-radiation in silicon (a) and in germanium (b).

The γ-radiation attenuation coefficient related to Compton scattering, is proportional to the "electron density" of the substance NZ.

Electron-positron pairs can be produced only if the γ-ray energy $h\nu$ exceeds the rest energy of these particles which is about 1.02 MeV. The law of conservation of energy and momentum requires that a third particle (usually an atomic nucleus) participates in the production of a pair. Almost the entire excess energy of the photon is converted into kinetic energy of the produced particles:

$$E_{e-} + E_{e+} = (h\nu - 1{,}02) \text{ MeV}.$$

At energies near the threshold energy the γ-ray absorption coefficient associated with pair production (ξ_{pair}) is proportional to the quantum energy:

$$\xi_{pair} \sim NZ^2 (h\nu - 2mc^2);$$

For higher energies this relationship is a logarithmic function:

$$\xi_{pair} \sim NZ^2 \ln (h\nu).$$

In light semiconductors, for instance in silicon, pair production greatly influences the total γ attenuation coefficient only if the radiation energy exceeds 10–15 MeV. Figure 38 shows the dependence of the absorption coefficient associated with the above three processes, on the radiation energy, for silicon (a) and germanium (b).

2. *Mean Ionization Energy ε in Semiconductors.*

In the most detailed investigation of ionization in gases, the total number of ion pairs N, produced as a result of slowing down of a particle, was proportional to the decrease in kinetic energy, ΔE, i.e., $N = \Delta E / \varepsilon$ where ε is the mean ionization energy. For gases, ε is much higher than the ionization potential and its value is usually about 30 eV/9/. The mean ionization energy is of great importance since it determines the possibility of counting and spectrometric measurement of particles and of high energy γ-rays by using ionization chambers and gas-filled counters. The value of ε for gases is not influenced by the initial energy and by the type of ionizing radiation.

We shall denote a similar parameter, characterizing ionization in semiconductors and in dielectrics, by ε. An experimental study of the ionization of solids by charged particles was undertaken in order to

112

develop "solid-state ionization chambers" or "solid counters"/159/ which were supposed to have certain advantages over gas-filled devices. These advantages included a high stopping power (thousands of times higher than that of gases) and small ε values in dielectrics and particularly in semiconductors.

Today semiconductor counters made of perfect silicon and germanium single crystals provide high energy resolution and are increasingly used in experimental and nuclear physics. The operation, design and technology of semiconductor counters have been discussed in detail/10,160/.

The main difficulty in the determination of the real (maximum) value of ε is to avoid (or correctly take into account) losses of non-equilibrium electrons and holes due to recombination, and the temporary capture of part of the carriers on local centers. Therefore, during the first experiments, which were carried out with natural diamonds and with alkali halide crystals there were considerable "losses associated with the freed charge". In 1951 MacKay carried out the first experiments during which almost all non-equilibrium carriers, freed by α-particles were collected. For his experiments MacKay used single germanium crystals with planar *p-n* junctions produced by doping during the crystal growth. A scheme of the experiment is shown in Figure 39.

If the *n*-type region has a positive potential with respect to the *p*-type region, i.e., if the *p-n* junction is reverse biased, then almost the entire potential drop will be on the *p-n* junction. The strongfield region depleted of carriers, can be considered an insulator separating two conductors. If the incident charged particles are slowed down by the strong field, then the excess electrons and holes, produced as a result of ionization, will move towards the *p-n* junction boundaries.

Excitation by thin "probing" light beams incident on the crystal parallel to the *p-n* junction in the strong-field region has shown that all carriers are collected in the region of the main absorption band where the quantum yield is equal to unity/13/. This means that recombination of electrons and holes (freed by the light) may be disregarded.

Those experiments, as the experiments on ionization by α-particles, were carried out in the region of "saturation currents" on the reverse branch of the volt-ampere characteristics of the *p-n* junction.

Figure 40 shows the equivalent circuit of a crystal with a *p-n* junction and an amplifier input. R_b and C_b are the resistance and the

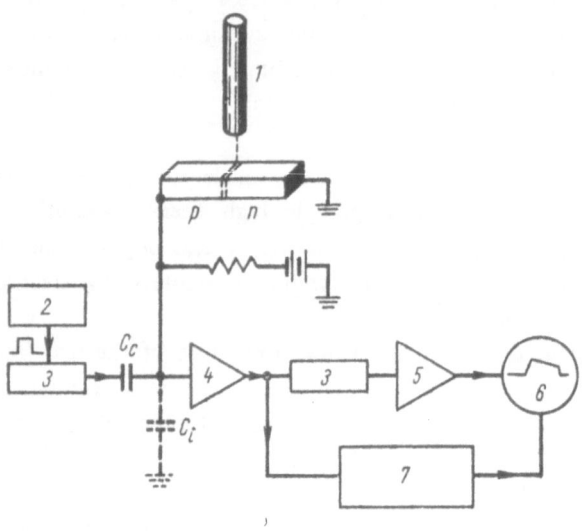

Figure 39. Schematic presentation of an experiment for the determination of the mean ionization energy by α-particles in semiconductor crystals with *p-n* junctions: 1—α-particle source; 2—generator; 3—attenuator; 4—preamplifier; 5-amplifier; 6—oscilloscope; 7—scanner.

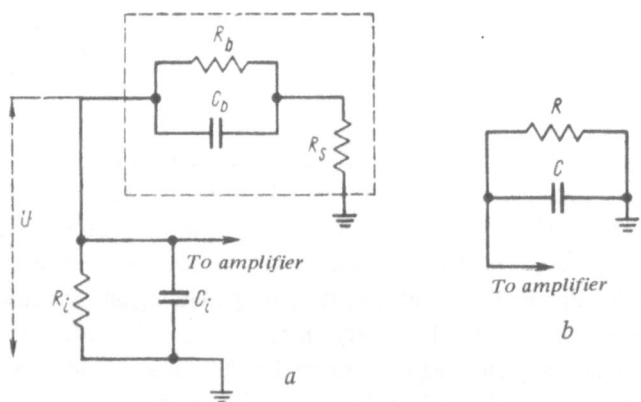

Figure 40. Equivalent circuit of a crystal with a *p-n* junction (*a*) and of the amplifier input (*b*).

capacitance of *p-n* junctions, R_i and C_i are the impedance components of the amplifier and R_s is the series resistance of the crystal bulk

114

and the contacts. If a reverse bias is applied to a crystal with a *p-n* junction, $R_b \gg R_s$ and therefore R_s can be disregarded, and the equivalent circuit becomes a simple parallel circuit RC (figure 40b) where $C = C_b + C_i$ and $R = R_b R_i/(R_b + R_i)$. The migration time of carriers through the strong-field region is considerably shorter than the time of RC relaxation and thus the result of ionization by α-particles is equivalent to the production of a current pulse through the barrier layer of the *p-n* junction involving a maximum change in potential $\Delta U = QC^{-1}$ where Q is the effective charge transferred through the strong field.

Figure 39 shows that the potential is applied from the crystal to the input of a wideband amplifier. The potential is applied to the specimen through a resistor equal to about R_b. The outlet signal of the preamplifier is shunted off to start the scanning of the oscillo-gram. The attenuator between the preamplifier and the main amplifier had a 0.17 μsec delay line (coaxial cable). The amplifying system had a 35 Mc band width and a maximum amplification of 100 decibel. The calibration circuit included a rectangular pulse generator with a rise time of 10^{-8} sec. The calibrated pulse produced a current pulse (amount of electricity) at the input circuit of the preamplifier, equivalent to the current pulse passing through the *p-n* junction as a result of ionization by α-particles. A comparison of the amplitudes of calibration pulses and of pulses produced by α-particles was used to determine the amount of electricity transported through the strong-field region in the *p-n* junction, by electrons and holes produced as a result of the stopping of a single α-particle.

The author of /161/ used a weak Po source with a collimator consisting of small coaxial apertures. The measurements were carried out in vacuo in order to avoid scattering of α-particles and energy losses in air. The maximum scatter of impingement points of α-particles on crystals was somewhat greater than the width of the strong-field region. As a result of this, there was a certain scattering of amplitudes since the diffusion of non-equilibrium carriers, produced outside the strong-field region was accompanied by recombination, before the carriers could be trapped by this field. The height of pulse scattering agreed with that calculated on the basis of the width of α-particle beams. The calculations were based on the maximum amplitude of pulses. Changing the voltage applied to the *p-n* junction, within the range of 0.3—15V, did not change the maximum amplitude of pulses.

115

According to MacKay the mean energy of pair production by α-particles is 3.0 ± 0.4 eV in silicon and 3.6 ± 0.3 eV in germanium. According to the same author the transfer time of carriers in the strong-field region does not exceed 10^{-10} sec. MacKay believes that the drift of non-equilibrium carriers and holes in the field starts immediately after their production.

In reality a state of "electron-hole plasma" exists over a certain time in the vicinity of the ionizing α-particle tracks. The polarization produces a space charge field which neutralizes the applied external field. In the plasma region which is not penetrated by the external field, the decrease in the concentration of electrons and holes is due to bipolar diffusion. The time needed for diffusion flow is usually short, and does not exceed $10^{-9} - 10^{-8}$ sec for ionization of silicon by α-particles. The rate of recombination in the plasma region can be quite high. Therefore, the energy per particle pair increases somewhat at high ionization densities and this must be taken into account when determining the energy of fission products by means of semiconductor counters /10/.

Soviet authors/162/, who have investigated ionization in such semiconductors as a result of the stopping of fast electrons, used the process of "collecting" non-equilibrium carriers which had been produced outside the strong field in crystals with p-n junctions. This was necessary because scattering of electrons in the bulk did not allow to use methods suitable for heavy particles.

An analysis of those works and a study of ionization in germanium by X-rays/163/ and γ-rays/164/ has shown that the mean ionization energy is almost unaffected by the nature of ionizing radiation.

If the experimental conditions are such that it is possible to use ionizing radiation pulses with a sharp front (for instance, pulsed beams of electrons or X-rays penetrating the crystal) then the mean ionization energy can be readily determined by the kinetic method used by S.M. Rivkin in a study of excitation by light /52/.

If it is possible to disregard the trapping of non-equilibrium carriers at local "capture levels" (for instance in pure germanium single crystals at room temperature), the kinetic method may be used to obtain accurate values of ε for excitation with fast electrons/165/. The value of ε was found to be 2.4 ± 0.2 eV, which is somewhat smaller than the value of ε for ionization by α-particles. The drawback of the kinetic method for the study of crystals in which carriers are captured by local centers, is that the obtained value of ε may be higher than the mimimum value, characterizing the energy distribution of absorbed radiation between the newly formed electrons and holes on one hand and the lattice vibrations (phonons) on the other.

Data on the ionization energies of semiconductors and diamond (E_g = 5.6 eV) at room temperature are given in Table 6.

Table 6. Forbidden band width E_g and mean ionization energy ε for semiconductors at room temperature. The value of $ε'_β$ was calculated /166/ by taking into account reflection of part of the electron beam.

Semiconductor	E_g	Plasmon energy	$ε_α$	$ε_β$	$ε_β$	$ε_γ$
ZnSb	0.165 /8/	12.7 /167/	—	—	—	1.2(?) /168/
Ge	0.665 /8/	15.5 /167/	2.85±±0.1 /169/	2.4±0.2 /165/	—	2.5±0.3 /169, 171/
Si	1.12 /8/	16.6 /167/	3.60±±0.05 /172,	4.0±0.2 /173/	3.6±0,2	3.55±±0.1 /174/
GaAs	1.35 /8/	15.5 /167/		6.3 /175/	4.8	—
GaP	2.35 /8/	16.6 /167/	7.8±0.8 /176/	—	—	—
CdS	2.4 /8/	—	7.2 /182/	9.3±1.0 /178/	—	—
SiC	2.86 /8/	—	—	9.0±0.7 /165/	—	—
PbO	3.0 /179/	—	—	—	—	8.0±1.2 /179/
C (Diamond)	5.6 /8/	31 /167/	—	18.5±±1.5 /180/	—	—

3. Energy Consumed in the Ionization and Excitation of Lattice Vibrations.

A simple model of impact ionization in semiconductors has been suggested by Shockley/181/. The model may be used to correlate data obtained by independent determinations of the quantum yield of photoionization in the ultraviolet region (which in germanium and silicon increases above photon energies equal to three times the forbidden gap) with values of ε obtained during the stopping of charged particles. Scatterings on lattices with the highest vibration frequencies (phonons) are most characteristic for charged carriers with a kinetic energy sufficient to produce impact (secondary) ionization. Such vibrations in diamond-type lattices are called "Raman vibrations" and correspond to the displacement of two face-centered sublattices in opposite directions.

The quantum energies of such vibrations E_R, determined on the basis of data on the scattering of "cold" neutrons, is equal to 0.063 eV for Si and to 0.037 eV for Ge. According to Shockley, all collisions of carriers with a lattice involve a loss of kinetic energy. The value of E_R is one of the parameters in Shockley's model. Other parameters used in this model are:

a. E_i — minimum (threshold) ionization energy calculated from the edge of the valence band; carriers with an energy $E > E_i$ can form secondary pairs;

b. L_R — mean free path of a carrier between two consecutive acts of scattering;

c. $r = L_i L_R$ where L_i is the mean free path of a carrier of energy $E > E_i$ between impact ionization acts.

Thus, an electron with an energy, $E > E_i$ produces an average of r phonons per single ionization act. The probability that a carrier, which moves in the crystal, would cause ionization before it loses the excess energy $(E - E_i)$ can be determined as follows. In order to lose the excess energy a carrier must undergo a collision entailing an emission of photons: $c = (E - E_i)/E_R$. Such a carrier passes a distance equal to cL_R. The probability that it will not cause ionization on its path is:

$$\exp\left(-\frac{cL_R}{L_i}\right) = \exp\left(-\frac{E - E_i}{rE_R}\right)$$

The probability of ionization $P(E)$ when the carrier is slowed down from an energy E to E_i will then be equal to:

$$P(E) = 1 - \exp\left(-\frac{E - E_i}{rE_R}\right)$$

Taking into account the experimental data on the quantum yield of Ge and Si photoionization Shockley assumed that E_i is equal to 0.68 eV for Ge and to 1.1 eV for Si. On the basis of the theories of ionization losses, he assumed that the energy "exchange" between primary electrons and holes is distributed between secondary impact ionizations and phonon generation. If the energy of carriers is reduced to values smaller than E_i, the residual energy E_F, is also consumed for the production of phonons. Thus, the mean energy per pair of non-equilibrium carriers is equal to:

$$\varepsilon = E_i + rE_R + 2E_F.$$

Using the values of r obtained from experimental data on photoionization and assuming $E_F = 0.6E_i$, Shockley obtained ε (Ge)=3.6 eV and ε (Si)=3.5 eV.

In contrast to Shockley, U.M. Popov/182/ assumed that for primary electrons and holes of energy E much greater than E_i, impact ionization is much more probable than scattering, so that the stopping of these carriers at the beginning of the process is determined by impact ionization alone and generation of phonons takes place only if $E < E_i$.

A very interesting theory on the relation between the mechanism of the ionization process and the plasma vibrations of valence electrons was proposed by Zaremba /170/ and analyzed and generalized by Klein/166/. Klein believes that the primary process of charged particles stopping is not associated with interactions between the moving particles field and individual electrons but with the interaction between the plasma and valence electrons which produces quanta (plasmons) whose energy is:

$$\hbar\omega_p = \hbar \left(\frac{4\pi Nq^2}{m}\right)^{1/2},$$

where N and m denote the concentration and mass of valence band electrons. Indeed, the values of "characteristic energy losses" from fast electrons in semiconductors with diamond-type lattices is close to the calculated value of $\hbar\omega_p$. The plasma vibrations are rapidly damped and the plasmon energy is transferred to one of the electrons which passes into the conduction band/183/. It is also possible that the plasmon energy is converted into "local heating" of electrons in the lattice within the limits of a localized volume of the crystal. Klein believes that the first kind of plasmon decay is more probable and lasts about 10^{-15} sec. The electron and hole pairs produced by this process have a high energy and are stopped as a result of impact ionization and phonon generation. The experimentally found dependence of ε on E_g has led to the derivation of an equation: $\varepsilon = 2.67 E_g + 0.87\,\mathrm{eV}$. This dependence is represented in Figure 41 by a continuous line /166/.

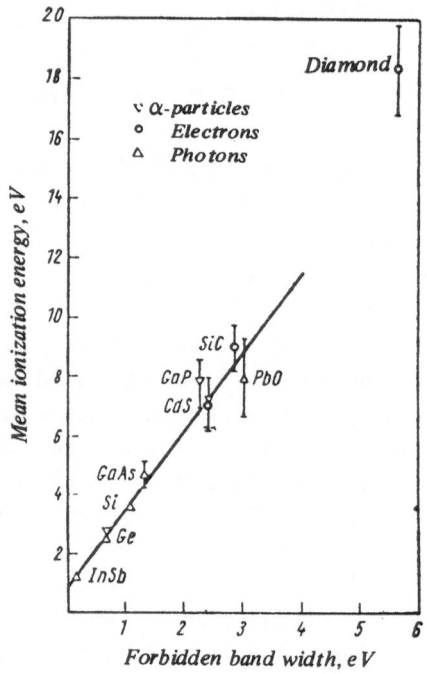

Figure 41. Relationship between the forbidden band width and the mean energy of formation of carrier pairs. Continuous line: $E = 2.67\, E_g + 0.86$ /166/. The experimental results were taken from Table 6.

It is still unclear whether excitons can be produced (together with carrier pairs) by charged particles passing through the crystal. It may be assumed that in germanium and silicon, in which the binding energy of excitons is very small, this process cannot have a marked effect on the value of ε. However, it may apparently lead to additional energy losses in crystals in which the binding energy of excitons is much greater than the value of kT, at the temperature of the experiment.

PART TWO

The production of semiconductor devices, which are widely used in electronics, became possible as a result of the considerable progress in the physics and production technology of very pure and perfect semiconductor single crystals. Negligible concentrations of impurities or slight structural imperfections impair the performance of semiconductor devices. Therefore, nuclear radiation, which causes marked structural damage in semiconductors, leads to considerable, mostly irreversible, changes in the properties of semiconductors which impair and often completely change their properties.

This necessitated an extensive study of the operation of semiconductor devices under various irradiation conditions and of the physical phenomena taking place as a result of semiconductor irradiation. The study of such phenomena is closely related to the progress in the radiation physics of semiconductors. However, the degree of radiation damage in semiconductor devices and the rate of its accumulation are determined not only by changes in the electro-physical properties of semiconductors but also by the service conditions, design and production of these devices. Part Two of this book is devoted chiefly to the elucidation of such problems.

The detrimental influence of radiation on semiconductor devices does not prevent their use. However, a detailed knowledge of the nature of radiation damage is required.

It has been found that the irreversible damage associated with changes in the electro-physical properties of semiconductors is approximately proportional to the radiation dose. Prolonged irradiation under conditions similar to those prevailing in the radiation belts of the earth leads to a gradual increase in damage. However, this does not lead to catastrophic failure of such devices. An exception is, perhaps, pulsed radiation, but there too a certain proportionality exists between the amount of radiation in the pulse and the magnitude of the produced effect. If the effect is slight and does not lead to complete failure of the device it can still conserve its full or partial service capability. The ability of irradiated semiconductors to retain their basic properties (even if there is a change in basic parameters) is called radiation stability.

The amount of radiation (integrated flux, exposure, or absorbed dose) which changes the basic properties of semiconductor devices beyond certain limits is taken as a measure of radiation stability. The limits are set as a certain maximum value of one of the basic parameters, and that value is usually determined by the operating conditions of the device; therefore, there are no universally accepted norms. Hence it is useful to relate the radiation stability criterion of a device to the physical properties on which its operation depends. Such a relation allows us to predict the radiation stability by analyzing the variation of certain parameters, which govern the subsequent increase in radiation damage.

Despite the great variety of existing devices (and those in design) the principles of operation of most of them are the same. Therefore it is useful to determine the influence of radiation (chiefly, neutron and gamma radiation) on transistors and diodes. Most of our attention will be devoted to residual, usually irreversible, changes but we shall also deal briefly with pulsed radiation.

We shall discuss radiation damage first in transistors and then in diodes since, despite the more complex structure of transistors, the processes taking place there under the influence of radiation are simpler and have been investigated in greater detail. Bulk damage in transistors is due mainly to a change in the lifetime of minority carriers, caused by radiation, and it is little influenced by a change in the conductivity of the material.

From this standpoint the processes taking place in diodes are more complex and therefore the theory of their operation under irradiation lags behind the experimental studies.

CHAPTER I

RADIATION EFFECTS IN TRANSISTORS

1. Experimental Determination of the Properties of Irradiated Transistors.

A study of semiconductor devices under irradiation has shown that most failures are due to considerable changes in the properties of the transistors. An example of this is the failure of the American communication satellite Telstar-1/184/. Studies of the operation of semiconductor devices under corpuscular and electromagnetic irradiation have shown that transistors have a low radiation stability.

An extensive study of radiation effects in transistors has shown that none of their characteristic parameters remains unchanged under irradiation. The input impedance and the feedback coefficient (h_{11} and h_{12}) of common-base circuits increase proportionally to the increase of the integrated flux of fast neutrons/185/. At the same time the output resistance (h_{22}) in the common-base circuits decreases monotonously /186, 187/. However, those changes are mostly determined by the same processes that change the basic properties of transistors during irradiation, such as the transport coefficient of the emitter current α and the base current B (the *d.c.* gain coefficient) and the reverse collector current I_{co}. Since these properties change more than other properties at a given level of irradiation they shall be given most of our attention.

For the study of radiation effects in transistors, it is convenient not to use absolute values, like $\alpha(1 - \alpha)$ or B, but relative values reduced to initial properties, i.e. α_Φ/α_0, $1 - \alpha_\Phi/(1 - \alpha_0)$ and B_Φ/B_0. Henceforth the subscript Φ shall designate the parameters of transistors and other semiconductor devices during and after irradiation while the subscript o will designate initial properties.

The results obtained in a study of irradiated transistors of different types and made of different semiconductor materials has shown that the value of $1 - \alpha_\Phi$ increases linearly (at least at the beginning of the irradiation) with increasing integrated flux or radiation dose. This value is closely associated both with α_0 and with the base current transport coefficient B_Φ which can be calculated (with an accuracy of unity) from: $B_\Phi = 1/(1 - \alpha_\Phi)$

Therefore, an increase in $(1 - \alpha_\Phi)$ is equivalent to a decrease in the coefficient B_Φ. Experimental data on the changes in $(1 - \alpha_\Phi)$ and

$1/B_\Phi$ are shown in Figures 42 and 43. These data were obtained by irradiating transistors with 2 MeV electrons, with Co^{60}-γ rays and with neutrons in a reactor.

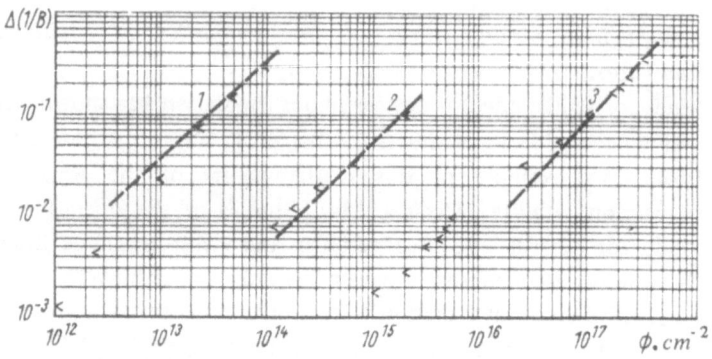

Figure 42. Dependence of the reciprocal of the base current transport coefficient on the integrated neutron flux (1), on the 2 MeV electrons flux (2) and on the flux of γ-rays from a Co-60 source (3) for low power *n-p-n* silicon (2N336) transistors prepared by pulling from a melt.

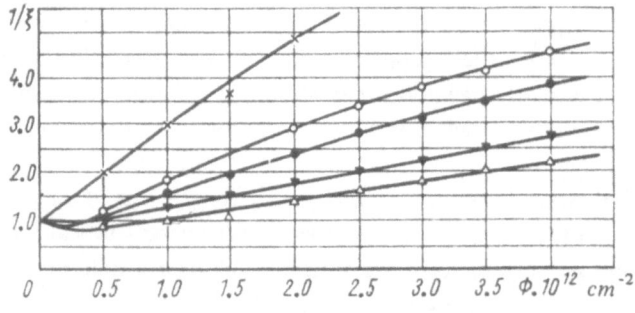

Figure 43. Dependence of $1/\xi$ on the integrated flux of fast neutrons. The experiments were carried out with low-power high-frequency germanium *p-n-p* transistors of various types.

The integrated flux of incident particles is plotted on the abscissa while the values of either $1/\xi = (1 - \alpha_\Phi)/(1 - \alpha_o)$, or $\Delta(1/B) = 1/B_\Phi + 1/B_o$ are plotted on the ordinate.

Etching of the surfaces of highly irradiated transistors has little effect on the base current transport coefficient. This has been confirmed by the data in Figure 44, which shows the dependence of $1/B$ on the emitter current before and after irradiation and also after etching the surface of one of the irradiated low-power n-p-n mesa transistors 2N706/188/. The transistors were irradiated in a "Triga" reactor with an integrated fast neutron flux of $8.8 \cdot 10^{13}$ cm^{-2}

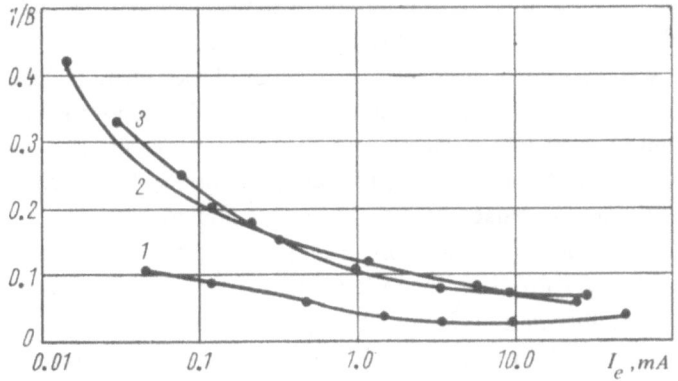

Figure 44. Dependence of $1/B$ on the emitter current in silicon n-p-n transistors (2N706) before irradiation (1), after irradiation with an integrated ($8.8 \cdot 10^{13}$ cm^{-2}) neutron flux (2) and after irradiation followed by surface etching (3).

This should not mean that the properties of semiconductor surfaces are not changed by ionizing radiation and that the parameters of the devices are not influenced by the radiation effects on the surface of semiconductors. During the initial stage of irradiation of high-frequency p-n-p germanium transistors the base current transport coefficient is determined chiefly by the radiation effects on the surface of the near-emitter region of the base (Figure 43). Those effects upset the linear dependence of $1/B_{\Phi}$ and $(1-\alpha_{\Phi})$ on the integrated flux or on the radiation dose. It has been found that the surface processes which lead to a change in the amplifying properties of transistors, have a maximum which is reached with relatively small radiation doses. This allows us to differentiate between surface effects and bulk effects since the latter are additive, at first approximation.

The magnitude and nature of the surface component of radiation effects on the current transport coefficients α and B depend on many factors, primarily on the initial surface state of the crystal. For some transistors this component is zero while for others it is several times the bulk component, at radiation doses corresponding to the maximum. It should be pointed out that for small radiation doses the surface component is larger than the bulk component in devices operated at the highest frequencies. In some cases the surface component leads to an increase in B during irradiation, but a decrease may also be caused. The physical processes taking place under the influence of radiation on the surface of semiconductors, which lead to the above phenomena, shall be discussed below.

As we have shown above the dependence of the radiation changes in the reciprocal of B or $(1-a_\Phi)$ on the integrated flux is not the same for all semiconductors in a given device. It has been found that the scatter in the values of $1/B_\Phi$ caused by bulk processes, is due to differences in the frequency characteristics of the semiconductors.

Data on germanium p-n-p or n-p-n alloyed transistors /189/ (see Fig. 45) show that a correlation exists between the rate of radiation-induced change in the emitter current transport coefficient and the limiting frequency f_α^* for that coefficient in various samples. A correlation also exists between the rate of radiation changes in α and the frequency characteristics of transistors of other types, including those with a heterogeneous base. For instance, (see the data in Fig..46/190/) in the case of high-frequency planar Si transistors (2N918) a linear relationship exists between the reciprocal of the increase in the emitter current transport coefficient $1/\Delta\alpha$ and the limiting frequency f_T at which the modulus of the base current transport coefficient $|B_f|$ is equal to unity.

The total change in the reverse collector current (I_{co}) under the influence of radiation also consists of two components: a bulk component and a surface component. Figure 47, which shows the dependence of $I_{c\Phi}/I_{co}$ on Φ for low-power alloy-type p-n-p germanium transistors, indicates that the bulk component is a linear function of the integrated neutron flux. The same is true for the bulk component of silicon devices /191/, although its origin is different from that in germanium transistors.

The surface component of the reverse collector current I_{co} is governed by ionization processes in the surrounding medium and on the surface of the crystal and depends both on the radiation dose and on the operating conditions of the transistor (as well as on other conditions).

---

* f_α is the frequency at which the modulus of α decreases by a factor of $\sqrt{2}$.

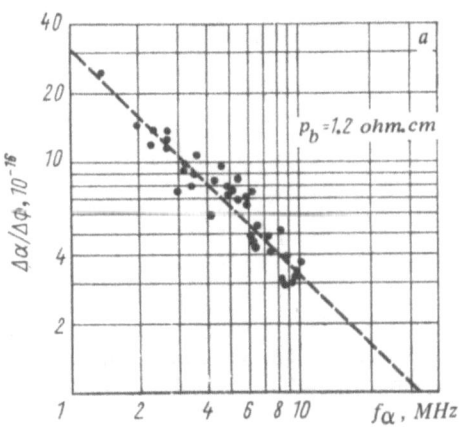

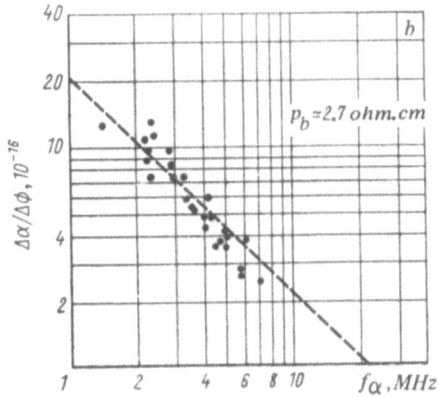

Figure 45. Dependence of the mean rate of decrease of the emitter current transport coefficient on the limiting value f_α for alloyed *p-n-p* (a) and *n-p-n* (b) germanium transistors.

2. Main Functions Determining the Rate of Change of Transistor Parameters as a Result of Bulk Processes. Criteria of Radiation Stability of Transistors.

Experimental data show that the decrease in the lifetime of minority carriers in the base region is responsible for the decrease in the current transport coefficients α and B as a result of changes in the bulk properties of the material caused by irradiation. An increase in the

127

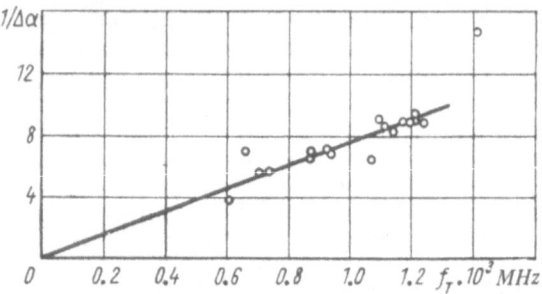

Figure 46. De, ndence of $1/\Delta\alpha$ on the limiting frequency for several planar silicon transistors (2N918).

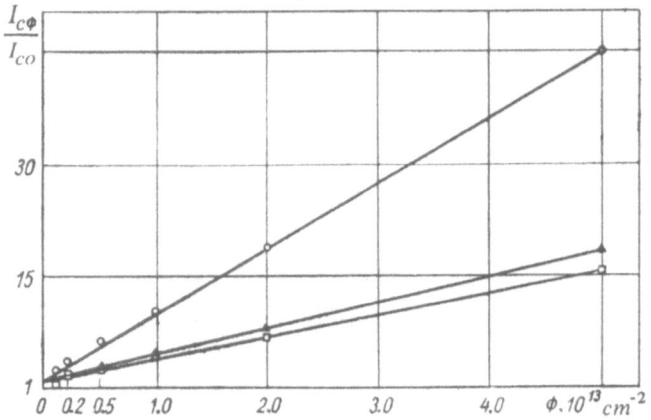

Figure 47. Dependence of the relative change in the reverse collector current on the integrated flux of fast neutrons for three alloyed *p-n-p* germanium transistors.

number of recombinations among carriers injected into the base decreases the number of carriers reaching the collector junction and of which the collector current is made up. The net result of this process is a decrease in the coefficient α.

The emitter current transport coefficient α can be presented as the product of three terms /192/: of the emitter efficiency γ characterizing its ability to inject minority carriers into the base region; of a transfer coefficient of minority carriers through the base β, which determines the recombination losses, and of the collector efficiency α_i, which

128

depends on the multiplication of charge carriers in the collector junction, as a result of impact ionization, and on the change in the conditions of diffusion of minority carriers in the collector region:

$$\alpha = \gamma\beta\alpha_i. \qquad (87)$$

The coefficient β represents the losses of minority carriers as a result of bulk recombinations and recombinations on the base surface adjacent to the emitter. The coefficient β can be, in turn, written as the product of two terms, one of which β_s represents surface processes and the second β_v represents recombinations in the bulk. Thus for α we have:

$$\alpha = \gamma\beta_v\beta_s\alpha_i. \qquad (88)$$

All terms in (88) are affected by radiation. However, the degree and rate of change are not the same. The bulk component of the minority carriers transport coefficient through the base β_v is greatly influenced by radiation. Therefore by taking all other terms constant we find that the bulk component of the base recombination current will be, on one hand, equal to:

$$I_{br} = I_e\, \gamma\beta_s\, (1 - \beta_v), \qquad (89)$$

and on the other hand:

$$I_{br} = q \int_0^{w_b} \frac{p\,(x)\,dx}{\tau\,(x)} = \frac{Q_b}{\tau_b}\,, \qquad (90)$$

where I_e and I_{br} are the emitter current and base recombination current respectively; Q_b is the charge accumulated in the base and τ_b is the mean value of the bulk lifetime of minority carriers averaged over the entire base:

$$\tau_b = \int_0^{w_b} p\,(x)\,dx \,\bigg/ \int_0^{w_b} \frac{p\,(x)\,dx}{\tau\,(x)}\,, \qquad (91)$$

129

where $p(x)$ is the concentration distribution of injected carriers.

Taking into account equations (89) and (90), equation (88) can be written as follows:

$$\alpha = \gamma\beta_s \left(1 - Q_b/I_e \, \gamma\beta_s\bar{\tau}_b\right) \alpha_i. \tag{92}$$

Since the ratio of the charge accumulated in the base Q_b to the current transported by the injected minority carriers is equal to the mean transit time t_m of carriers from the emitter to the collector /193/, equation (92) can be written as follows:

$$\alpha = \gamma\beta_s\alpha_i \left(1 - t_m \, /\bar{\tau}_b\right). \tag{93}$$

This equation is valid for both diffused and drifted transistors.

The lifetime of minority carriers in semiconductors is most sensitive to radiation. Usually, this lifetime is greatly changed by radiation doses which only slightly affect other properties (for instance, conductivity). Under such conditions the dependence of the lifetime on the integrated radiation flux Φ can be written as follows:

$$1/\tau = 1/\tau_0/ + K\Phi, \tag{94}$$

The above equation has been experimentally verified for a wide range of τ values. In (94) τ_0 is the initial lifetime and K is a proportionality coefficient which indicates the rate of lifetime changes upon irradiation (lifetime variation coefficient).

As we shall see below the value of K is influenced by many factors, among them by the concentration of majority carriers and by the injection level of minority carriers. For semiconductors with a nonuniform resistivity, such as the bases of drifted transistors, equation (94) is valid only for a point or plane with a given $\vec{K}$ value. However, if we introduce a coefficient $\bar{K}$ representing the mean value with respect to a certain volume, plane or direction, then (94) would be valid for any possible case. The coefficient $\bar{K}$ can be written:

$$\bar{K} = \frac{\iiint K(x, y, z) \, p(x, y, z) \, dx \, dy \, dz}{\iiint p(x, y, z) \, dx \, dy \, dz}. \tag{95}$$

Equation (94) shows that the lifetime of minority carriers decreases monotonously during irradiation. According to (93) this decrease is responsible for the fall of the current transport coefficient α or, to be more exact, of the bulk component of the transport coefficient of minority carriers through the base. This effect exists, to different degrees, in transistors of all types.

Taking into account equation (94) the equations for α_Φ and $(1-\alpha_\Phi)$ can be written as follows:

$$\alpha_\Phi = \gamma \beta_s \alpha_i \left(1 - t_{tr}/\overline{\tau}_{b0} - t_{tr} K\Phi\right), \qquad (96)$$

but since, usually, $\alpha_0 \approx 1$, we would have:

$$1 - \alpha_\Phi = 1 - \alpha_0 + t_{tr} \overline{K}\Phi. \qquad (97)$$

Those equations show a linear dependence of $(1-\alpha_\Phi)$ on the integrated radiation flux Φ. The above experimental results indicate that the linear dependence of radiation-induced changes exists only over a certain range of radiation fluxes. Outside that range the changes become gradually slower and the function $(1-\alpha_\Phi)=f(\Phi)$ ceases to be linear as it approaches a constant value. This is due to the fact that in reality both K and the transit time t_{tr} are changed by radiation. The transit time is determined not only by the dimensions of the base region but also by the lifetime of minority carriers in that region. However, since at lifetimes τ much longer than t_{tr}, this dependence has little influence, it is usually disregarded, and t_{tr} is determined either as a function of the dimensions of the transistor or of its frequency characteristics, as follows::

a) For alloyed transistors with a homogenous base, t_{tr} is determined as a function of the base thickness w_b or of the threshold frequency of the emitter current transport coefficient f_α /194/.

$$t_{tr} = \frac{w_b^2}{2D_b} = \frac{2,43}{4\pi f_\alpha} \qquad (98)$$

where D_b is the diffusion coefficient of minority carriers.

b-t_{tr} is determined as a function of the threshold frequency $f = |B_f| f$ for which the modulus of the base current transport coefficient $|B_f|$

equals unity /195/:

$$t_{tr} = \frac{1}{2\pi f_T} - \frac{kT}{qI_e}(C_{cb} + C_e).$$ (99)

c-t_{tr} is determined as a function of the threshold frequency of the base current transport coefficient, for which the modulus is smaller by a factor of $\sqrt{2}$:

$$t_{tr} = \frac{1 - \alpha_0}{2\pi\alpha_0 f_B}.$$ (100)

The general equation for $(1 - \alpha_\Phi)$ can be written as follows:

$$1 - \alpha_\Phi = (1 - \alpha_0)(1 + A\overline{K}\Phi),$$ (101)

where (according to the various theories on t_{tr} and provided that $\alpha_0 \approx 1$)

$$A = \frac{0,195 B_0}{f_\alpha},$$ (101a)

or:

$$A = \frac{B_0}{2\pi f_T} \quad \text{at} \quad f_T \ll \frac{qI_e}{2\pi kT (C_{cb} + C_{eb})}$$ (101b)

or:

$$A = 1/2\pi f_B.$$ (101c)

The relative base current transport coefficient during irradiation, ξ depends on factors related to transistor design (A), base material (K) and the integrated radiation flux Φ

$$\xi = \frac{1 + B_\Phi}{1 + B_0} = \frac{1 - \alpha_0}{1 - \alpha_\Phi} = \frac{1}{1 + A\overline{K}\Phi}.$$ (102)

A graph for the above relation is shown in Figure 48. The integrated radiation flux is plotted (in Φ/AK units and on a logarithmic scale), on the abscissa and ξ on the ordinate. Equation (102) represents the dependence of the relative base current transport coefficient on the integrated radiation flux for transistors of all types (Figure 48). This equation may be plotted as two straight lines with a break at $\Phi_a = 0.135/A\overline{K}$ This integrated flux, known as the absolute stability flux /196/, is a useful criterion for the assessment of the radiation stability of transistors. At radiation fluxes not exceeding Φ_a it is usually possible to disregard the slight decrease (smaller than 11.5%) of the base current transport coefficient. For integrated fluxes exceeding Φ_a, the function $B_\Phi = f(\Phi)$ can be written as follows:

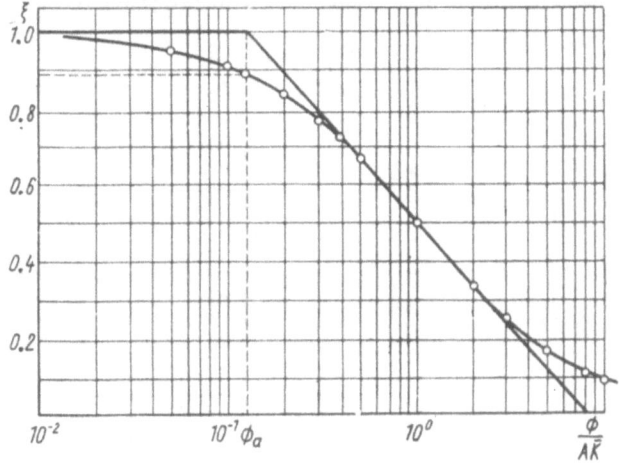

Figure 48. Dependence of the relative base current transport coefficient of transistors ξ on the integrated radiation flux reduced to the value $\xi = 0.5$

$$B_\Phi = B_0\,(1 - 0.575\,\lg \Phi/\Phi_a). \qquad (103)$$

The absolute stability flux for transistors Φ_a is determined by the frequency properties and can therefore be predicted on the basis of the lifetime coefficient $\overline{K}$ of the base material

$$\Phi_a = 0.85 f_B/\overline{K} = 0.69 f_a/B_0\overline{K} = 0.85 f_T/B_0\overline{K}. \qquad (104)$$

The above equation shows that the radiation stability is a linear function of both the frequency characteristics of semiconductors and of $\overline{K}$.

133

In order to predict radiation damage in the most common types of low-power germanium and silicon transistors it is necessary to determine the relationship between the relative base current transport coefficient and the integrated neutron flux. The flux corresponding to the absolute stability Φ_a for each device was found by measuring the threshold frequency of the base current transport coefficient f_B and using the mean value of $\overline{K}$ for a given type of transistor. After that, the transistors were irradiated and their base current transport coefficient was measured periodically. The agreement between experimental values and those calculated using the $\xi=F(\Phi)$ equation, is illustrated by Figure 49.

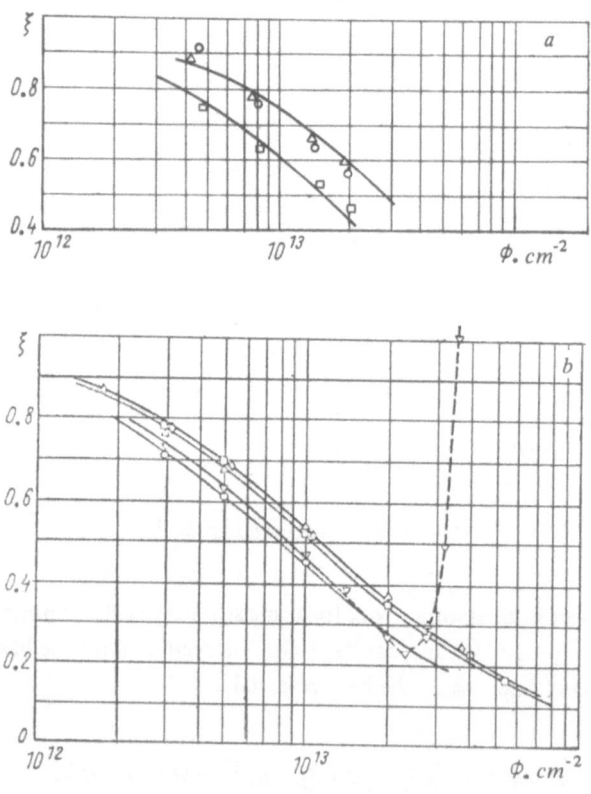

Figure 49. A comparison between the calculated (continuous current transport coefficient ξ on the integrated fast neutron flux same transistor type were tested): a—alloyed-diffused high-frequency c—high-frequency silicon diffused *n-p-n* transistors; d—alloyed

The value of the base current transport coefficient (B_{Φ}) after irradiation with a given fast neutron integrated flux may also be found with the aid of Fig. 50 /197/.

3. Behavior of Transistors Irradiated with Large Doses.

The equations showing the effect of radiation on the gain parameters of transistors, are not accurate for large radiation doses, because of bulk processes taking place under these conditions. It has been found that the deviation from linearity in the increase of $1/\xi$ with increasing integrated radiation flux becomes very marked at radiation levels for which the relative base current transport coefficient ξ is smaller than $0.1-0.2$. One of the main reasions for the disturbance of linearity is the change in transit time of minority carriers through the base (t_{tr}) upon irradiation. This change could be due to a change in the drift mobility of carriers, or to a change in diffusion conditions as a

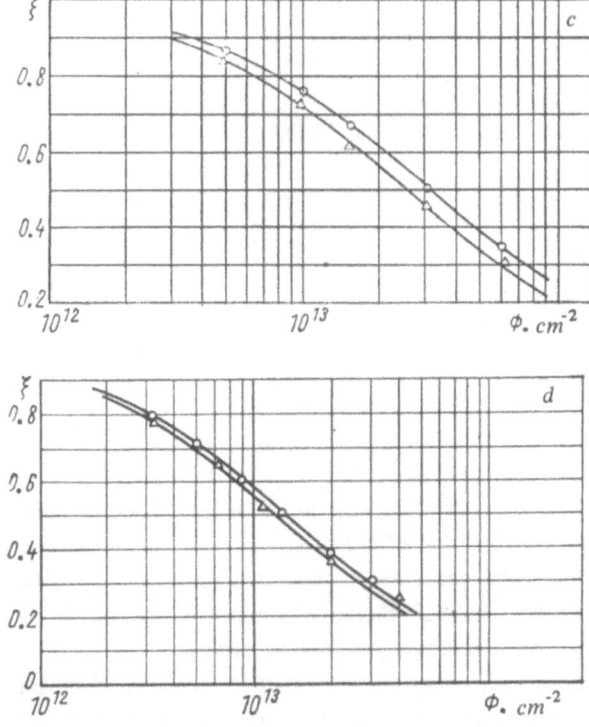

Figure 49. cont.

lines) and experimental (points) dependence of the relative base for the following low-power transistors (several specimens of the germanium transistors (p-n-p); b–alloyed germanium p-n-p transistors; germanium n-p-n transistors.

135

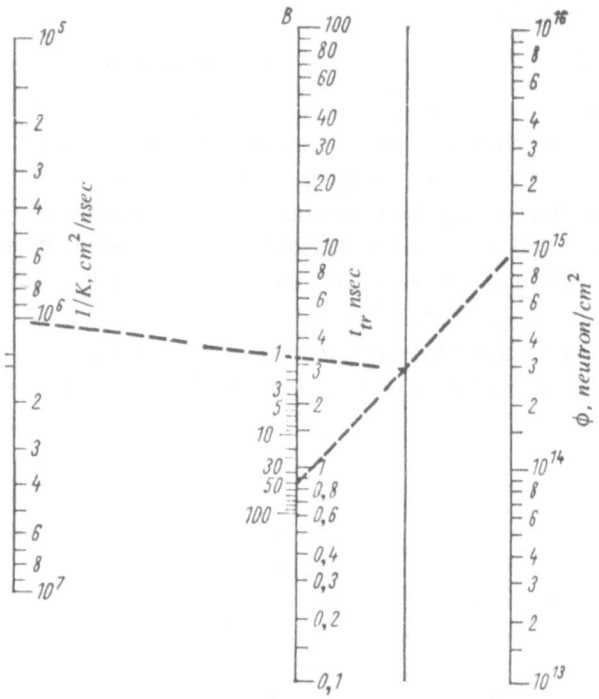

Figure 50. Nomograph for the determination of B from the integrated fast neutron flux Φ.

result of a considerable decrease in the volume lifetime and a decrease in the effective thickness of the base.

Since radiation produces various structural defects in the bulk of semiconductors, the drift mobility of minority carriers can be expected to decrease with increasing radiation dose (because of the increased probability of scattering on lattice imperfections). However, Gossick /86/ found that during neutron irradiation of n-type germanium, which produces large defect clusters, the drift velocity may increase with increasing integrated flux.

Closser /92/ found that in neutron-irradiated n-type germanium (ρ = 5 ohm·cm) the carrier mobility rapidly increase above an integrated flux of 10^{12} cm^{-2}. At Φ= 1.6·10^{12} cm^{-2} the mobility reaches a maximum which exceeds the starting carrier mobility by about 25% but afterwards it decreases rapidly and at Φ = 3.2·10^{12} cm^{-2} the carrier mobility is only 90% of the initial value.

Closser's data confirm experimentally Gossick's theory about the increase in the drift mobility of minority carriers at a certain stage of neutron irradiation. However, from the standpoint of transistor operation this increase is of little importance. The subsequent fast decrease in drift mobility is more important.

The available experimental data on the drift mobility of carriers are not sufficient to explain the changes during operation of irradiated transistors, particularly since those changes are much smaller than the change in lifetime τ. Therefore, the drift mobility is usually taken as constant, at least for integrated fast neutron fluxes up to 10^{14} cm^{-2}. The validity of such a simplification is confirmed by the satisfactory agreement between experimental results and theoretical calculations which do not take into account the influence of radiation on the drift mobility of minority carriers.

The effect of the decreased lifetime (upon irradiation) on the diffusion of carriers in the base and on their transit time can be conveniently demonstrated for transistors with a homogenous base. An exact equation for the bulk component of the carrier transport coefficient through the base of such transistors can be written as follows:

$$\beta_v = \text{sch} \frac{w_b}{L_b} \tag{105}$$

where w_b is the base thickness and L_b is the diffusion length of minority carriers in the base. Since usually w_b is much smaller than L_b the hyperbolic secant can be approximated by the first two terms of the expansion into series:

$$\beta_v = 1 - \frac{1}{2} \left(\frac{w_b}{L_b}\right)^2 \tag{106}$$

The hyperbolic secant cannot be represented by the two first terms of the series if the radiation decrease in minority carriers lifetime makes the diffusion length comparable with w_b. If we include the third term of the series sch w_b/L_b, the expression for β_v would read:

$$\beta_v = 1 - \frac{1/2 \left(\dfrac{w_b^2}{D_b}\right)}{\tau_b}\left[1 - \frac{5}{6}\cdot\frac{1/2\left(\dfrac{w_b^2}{D_b}\right)}{\tau}\right] \tag{107}$$

Taking the dependence of τ on Φ and using $\tfrac{1}{2}(w_b^2/D_b) = t_{tr.o}$ we have:

$$\beta_v = 1 - t_{tr.o}\left(\frac{1}{\tau_{bo}}+\overline{K}\Phi\right)\left[1-\frac{5}{6}t_{tr.o}\left(\frac{1}{\tau_{bo}}+\overline{K}\Phi\right)\right]. \tag{108}$$

A stricter derivation of (108) yields a value of 2/3 instead of 5/6 for the numerical coefficient. If we take into account $t_{tr\,o}/\tau_{bo} \ll 1$, we obtain:

$$\beta_v = 1 - \left(\frac{1}{\tau_{bo}}+\overline{K}\Phi\right)t_{tr.o}\left(1-\frac{2}{3}t_{tr.o}\overline{K}\Phi\right) \tag{109}$$

instead of:

$$\beta_v = 1 - t_{tr}\left(\frac{1}{\tau_{bo}}+\overline{K}\Phi\right)$$

Hence the transit time t_{tr} decreases upon irradiation. The rate of this decrease, i.e., $dt_{tr}/d\Phi = -2/3t^2{}_{tr.o}\,\overline{K}$, becomes more pronounced as its initial value is increased. Therefore, the effect of the decrease in t_{tr} due to the decrease in τ is most pronounced in low-frequency transistors and is negligible in high-frequency transistors.

According to equation (100) the product of B and the threshold frequency of this coefficient, is inversely proportional to the transit time t_{tr}. Therefore, the above product may serve as a measure of the change in t_{tr} for irradiated transistors. Table 7 shows values of that product for several low-power alloyed p-n-p and n-p-n germanium transistors before and after irradiation with fast neutrons.

As can be seen, radiation increased the value of $B \cdot f_B$ by a factor of 1.6, i.e., the transit time of carriers in the base decreased as a result of irradiation. Because of the concentration gradient of carriers in the base and of the field produced by that gradient, the drift component of the transit velocity of carriers exerts considerable influence on alloyed diffused germanium transistors and on diffused silicon transistors. Therefore, the decrease in the lifetime of carriers injected into the base has almost no influence on the transit time of carriers in such transistors. The product $B \cdot f_B$ for such transistors decreases monotonously during irradiation.

Table 7. Changes in $B \cdot f_B$ of alloyed Ge transistors upon irradiation

Transistor type	$B_0 \cdot f_{B0}$, KHz	$\dfrac{\Phi}{10^{14}}$, n/cm^2	$B_\Phi \cdot f_{B\Phi}$, KHz	$\dfrac{B_\Phi \cdot f_{B\Phi}}{B_0 \cdot f_{B0}}$
$p - n - p$	2520	1	4150	1.65
	5170	1	9150	1.77
	8750	1	13370	1.53
	2010	1.2	2940	1.46
$n - p - n$	785	0.65	1340	1.70
	1080	0.65	1780	1.65
	1100	0.65	1710	1.55

A decrease in the effective thickness of the base w_b causes a change in the transit time t_{tr} in high-frequency transistors with a homogenous base. The resistivity of such transistors is increased by irradiation. The effective base thickness of transistors is smaller than the real thickness. If we disregard the very narrow emitter p-n junction, this thickness would be:

$$w_b = w_{tech} - w_{pnk}(\Phi), \qquad (110)$$

where w_{tech} is the design thickness of the base and w_{pnk} is the thickness of the space charge layer which is determined by the potential applied to the collector junction and by the resistivity of the base material. The value of w_{pnk} increases with decreasing concentration of majority carriers in the base material. This increase can be determined by measuring the collector impedance.

The dependence of the majority carrier concentration on the integrated radiation flux (provided that the flux is relatively constant) would be as follows:

$$n_n(\Phi) = n_{n0} - K_\rho \Phi, \qquad (111)$$

where n_{n0} is the initial majority carriers concentration and K_ρ is a coefficient characterizing the rate of removal of those carriers.

By using an equation representing the width of the sharp p-n junction /129/ and equation (111) we obtain:

$$w_{pnk}(\Phi) = \sqrt{\frac{2\varepsilon\varepsilon_0(\psi_k - U)}{q(n_{n0} - K_\rho\Phi)}}, \tag{112}$$

and since (for such integrated fluxes) $K_\rho\Phi \ll n_{n0}$, we can write with adequate accuracy:

$$w_{pnk}(\Phi) = w_{pnk0}\left(1 + \frac{1}{2}K_\rho\Phi/n_{n0}\right), \tag{113}$$

where w_{pnk0} is the width of the p-n junction at $\Phi = 0$.

If we substitute the expression for t_{tr} from (98) into the equation for the minority carriers transport coefficient β_v, and take into account equation (113) we obtain:

$$\beta_v = 1 - \frac{1}{2}\cdot\frac{w_b^2}{D_b\tau_b}\left(1 - \frac{w_{pnk0}K_\rho\Phi}{w_b n_{n0}} + \tau_b\bar{K}\,\Phi - \frac{\tau_b w_{pnk0}\bar{K}K_\rho\Phi^2}{w_b n_{n0}}\right) \tag{114}$$

Let us analyse the above expression. Its derivative $d\beta_v/d\Phi$ is equal to:

$$\frac{d\beta_v}{d\Phi} = -\frac{1}{2}\cdot\frac{w_b^2}{D_b\tau_b}\left(\tau_b\bar{K} - \frac{w_{pnk0}K_\rho}{w_b n_{n0}} - \frac{2\tau_b w_{pnk0}K K_\rho\Phi}{w_b n_{n0}}\right)\cdot \tag{115}$$

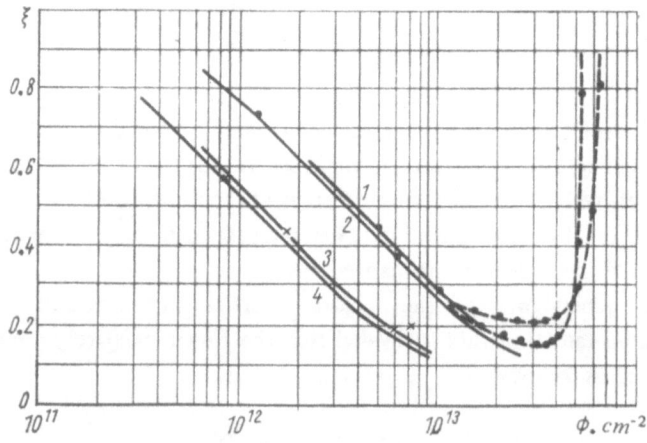

Figure 51. Dependence of the relative base current transport coefficient on the integrated fast neutrons flux in germanium p-n-p high-frequency (1 and 2) and low-frequency (3 and 4) high-power transistors ($P_{calc} = 10$ w).

At the beginning of irradiation, i.e., at $\Phi \approx 0$, this derivative is negative since as a rule $\tau_b K \gg w_{pnko}\bar{K}_p/w_b n_0$ and therefore the coefficient β_v decreases linearly during irradiation. However as the integrated flux increases the rate of the decrease of β_v as a result of irradiation diminishes gradually and becomes equal to zero when:

$$\Phi = \frac{w_b n_0}{2 w_{pnko} K_\mu} - \frac{1}{\tau_b \bar{K}} \tag{116}$$

Further irradiation increases β_v with a corresponding increase in B. The broadening of the collector junction, which is exhibited as an apparent recovery of the amplifying properties of the transistor, is terminated by a "puncturing" of the base, i.e. by short-circuiting the emitter and collector junctions. The experimentally found $\xi = F(\Phi)$ functions for alloy-type germanium p-n-p transistors, which illustrate their abnormal behavior under intensive irradiation, are shown in Figure 49 (low-power transistors) and in Figure 51 (high-power transistors), for various threshold frequencies f_α.

4. Effect of Recombination at p-n Junctions on the Current Transport Coefficient of Transistors.

Frequently we must take into account the influence of penetrating radiation on the emitter current transport coefficient α. In such a case there is a monotonous decrease in the effectiveness of the emitter, which acts as an injector of minority carriers into the base. This decrease is due to an increase in the rate of carrier recombinations in the p-n emitter junction. At low operation currents the influence of this phenomenon on the gain parameters of transistors increases and may even become predominant.

Sah et al. /198/ have shown that the total current passing through a junction can be represented as a sum of several components: electron current, hole current and recombination current (I_{rec}). Therefore the effectiveness of an emitter in a p-n-p transistor can be written:

$$\gamma = \frac{I_p}{I_p + I_n + I_{\overline{rec}}} = \frac{1}{1 + \dfrac{I_n + I_{rec}}{I_p}}, \tag{117}$$

where I_p- is the hole and I_n- is the electron component of the total emitter current:

$$I_p = \frac{qp_n D_p}{w_b}\left(e^{\frac{qU_{eb}}{kT}} - 1\right) \qquad (118)$$

and

$$I_n = \frac{qn_p D_n}{L_n}\left(e^{\frac{qU_{eb}}{kT}} - 1\right). \qquad (119)$$

It has been found /198/ that the bulk recombination component of the total current flowing through a p-n junction in the forward direction can be found from:

$$I_{rec} = \frac{qn_i w_{pn} A_{pn}}{\sqrt{\tau_{n0}\tau_{p0}}} \cdot \frac{2 \, \mathrm{sh}\left(\frac{qU_{eb}}{2kT}\right)}{\frac{q}{kT}(\psi_k - U_{eb})} f(b), \qquad (120)$$

where

$$f(b) = \int_0^\infty \frac{dz}{z^2 + 2bz + 1} \quad \text{and}$$

$$b = e^{-\frac{qU_{eb}}{2kT}} \, \mathrm{ch}\left[\frac{(E_t - E_i)}{kT} + \frac{1}{2}\ln\frac{\tau_{p0}}{\tau_{n0}}\right]$$

Here E_i denotes the Fermi level of a material with an intrinsic conductivity; E_t is the energy level of traps for recombination-generation centers; A_{pn} is the surface area of the p-n junction; ψ_k is the diffusion potential of the p-n junction; t_{no}, t_{po} are the lifetimes of electrons and holes respectively, in highly doped materials.

Since the electronic component of the current is very small we can disregard the term I_{rec} in equation (117) and obtain:

$$\gamma = \left[1 + \frac{\frac{2qn_i w_{pn}\,\mathrm{sh}\,(qU_{eb}/2kT)}{\sqrt{\tau_{n0}\tau_{p0}}\,q/kT\,(\psi_k - U_{eb})}f(b)}{qp_n D_p/w_b(e^{qU_{eb}/kT} - 1)}\right]^{-1} \qquad (121)$$

Since the numerator of the second term is proportional to exp $(qU_{eb}/2kT)$ and its denominator is proportional to exp (qU_{eb}/kT) an increase in the potential U_{eb} applied to the junction will decrease this term to zero and the effectivity of the emitter γ to unity.

Irradiation monotonously reduces the lifetime of carriers in the *p-n* junction $\tau_{pn} = \sqrt{\tau_{no}\tau_{po}}$, which approaches a very small value. As a result the value of the second term in (121) equals approximately unity at a potential U_{eb} and can be disregarded before irradiation is started.

Data on the emitter current transport coefficients $\alpha = F(I_e)$ before and after irradiation of alloy type germanium OS-44 transistors with various integrated neutron fluxes are shown in Figure 52 /199/; the data were obtained experimentally or calculated from (121). Since at room temperature the intrinsic carriers concentration in silicon is smaller than in germanium by a factor of $1.5 \cdot 10^3$, it can be expected that similar effects will take place in silicon transistors at emitter current densities 3 orders of magnitude higher than those in Figure 52.

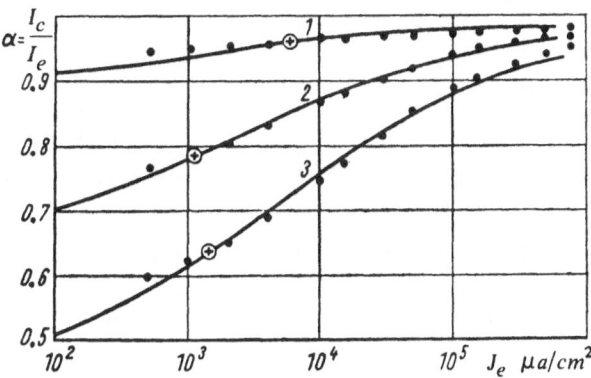

Figure 52. Calculated (continuous curves) and experimental (points) data on the relation between α and the emitter current density before (1) and after irradiation with $1.5 \cdot 10^{12}$ neutrons cm^{-2} (2) and with $3.8 \cdot 10^{12}$ neutrons cm^{-2} (3) of OS-44 *p-n-p* germanium transistors. Those functions "join" at points designated by $\oplus$.

The equation for I_{rec} (120) contains several terms which are an exponential function of the potential across the junction, U_{eb}, but have different values of the proportionality coefficients. Therefore the experimental dependence can be written, in general:

$$I_{rec} = I_{rec.0} \, \exp \, (qU_{eb}/mkT), \qquad (122)$$

where *m* is not equal to 2, as could be expected from (120), but has a value between 1 and 2 /200/.

Goben /201/ and Beicker /202/ found that in neutron irradiated silicon transistors the increase in the bulk component of the recombination current in the emitter p-n junction at relatively low current densities $(10^{-3} - 10^{-1}$ amp/cm^2) is almost totally determined by changes (in the emitter current coefficient), caused by radiation. In that case $m \approx 1.5$ (Table 8).

Table 8. Experimental values of m and K_I for some commercial transistors.

Transistor	Transistor type	Structure	m	K_I, 10^{22} A/cm^2/ neutron/cm^2
2N914	Planar,	$n-p-n$	1.5	3.3
SF2523	epitaxial		1.5	3.3
SF2524			1.5	3.3
A2409			1.52	4.7
2N718A	Planar	$n-p-n$	1.47	6.0
2N1613			1.47	6.0
2N498	Mesa-transistor	$n-p-n$	1.45	6.4
2N336	Grown from	$n-p-n$	1.56	5.5
2N338	a melt		1.64	5.5
2N1989	Diffused	$n-p-n$	1.38	6.8
2N327A	Alloyed	$p-n-p$	1.42	—

If we disregard the changes in carrier mobility during irradiation and the changes in their equilibrium concentration, then I_{rec} and γ in (120) and (121) would have only one parameter that depends on the integrated radiation flux. This would be the carriers lifetime in the p-n junction τ_{pn}, whose reciprocal increases linearly with increasing integrated neutron flux Φ /199/. Therefore the current I_{rec} can be calculated from:

$$I_{rec} = I_{rec}(0) + Q_{rec} A_{pn} K_{pn} \Phi \exp\left(\frac{qU_{eb}}{mkT}\right) \qquad (123)$$

or simply:

$$I_{rec} = K_I A_{pn} \Phi \exp \left(\frac{qU_{eb}}{mkT} \right) . \tag{124}$$

Here Q_{rec} represents the product of all terms in (120) that do not depend on Φ or on U_{eb}. It has also been assumed that:

$$1/\tau_{pn} = 1/\tau_{pn0} + K_{pn}\Phi.$$

The coefficient K_I in (124) equals a few units ($\cdot 10^{-22}$) of $(A/cm^2)/(neutron/cm^2)$. The mean values of this coefficient for several types of transistors are given in Table 8.

The mean coefficients for all types of transistors except for alloyed p-n-p irradiated with fast > 0.01 MeV neutrons are: $m = 1.49$, $(K_I) = 5.1 \cdot 10^{-22}$ $(A/cm^2)/(neutron/cm^2)$.

The fact that I_{rec} is a linear function of Φ indicates that the effectiveness of the emitter, at least during the initial period of irradiation (when this effectiveness is close to unity), would decrease linearly with increasing integrated flux Φ. Indeed, by using Γ to designate in the second term the product of parameters which are not changed by irradiation (in (121)) we obtain:

$$\gamma = [1 + \Gamma/\tau_{pn}]^{-1} = [1 + \Gamma/\tau_{pn0} + \Gamma K_{pn}\Phi]^{-1} \tag{125}$$

or

$$\gamma \approx \gamma_0 - \Gamma K_{pn} \Phi. \tag{126}$$

Thus, it is obvious that the mechanism of degradation of the emitter current transport coefficient α is related to Φ in the same way as the main process of recombination of non-equilibrium carriers in the base. Therefore, it is often difficult to differentiate experimentally between those effects, particularly in the range of current densities at which γ is close to unity. This is one of the reasons why a study of the effects of transistor irradiation yields erroneous data on the coefficient for the radiation-induced increase in the lifetime of minority carriers K in the base. In such a case the radiation stabilities of various transistors can be compared on the basis of the coefficient K' which is representative for the transistor as a whole.

5. Dependence of Radiation-Induced Changes in the Lifetime of Minority Carriers on the Electrophysical Properties of Irradiated Semiconductor Materials.

The coefficient K for the radiation-induced changes in the lifetime of minority carriers, a term in all equations used for the determination of the rate and degree of degradation of transistor parameters, is a convenient measure of the influence of nuclear radiation on the recombination properties of semiconductor materials. As we mentioned above the coefficient K can be considered constant over the range of integrated radiation fluxes that are of practical importance. The lifetime of carriers in irradiated semiconductors can be found from:

$$1/\tau = 1/\tau_0 + K\Phi. \tag{127}$$

The above equation has been confirmed by numerous experiments on the irradiation of semiconductors and semiconductor devices, whose parameters could be used to determine (directly or indirectly) the changes in τ /203/.

Figure 53 shows the dependence of τ on the integrated flux of fission-spectrum fast neutrons, for silicon specimens with p-type conductivity and various resistivities (several tens of ohms/cm) and for one specimen with n-type conductivity and a resistivity of several hundred ohms/cm /204/. The lifetime of irradiated specimens was determined from the rate of decrease in photoconductivity and from the steady-state photoconductivity. Similar results were obtained, in this work, with silicon diodes with p-type conductivity.

A linear relationship between $1/\tau$ and Φ was also observed in germanium irradiated with neutrons /204-207/ or γ-rays /208-210/.

Nevertheless the linear relationship /127/ may be disturbed at high irradiation doses. This has been confirmed by Inushi and Matsuura /211/ who irradiated oxygen-saturated silicon with γ-rays (Figure 54) and by Streetman /212/ who irradiated germanium with γ-rays.

The value of K depends on many factors, including the rate of implantation of recombination centers in the forbidden gap, the recombination properties of those centers (e.g. the electron and hole capture cross-sections) and their effectiveness, i.e., the level of electron filling, which depends on the position of the Fermi level with respect to the energy levels of those centers. In general the coefficient K may be written as follows:

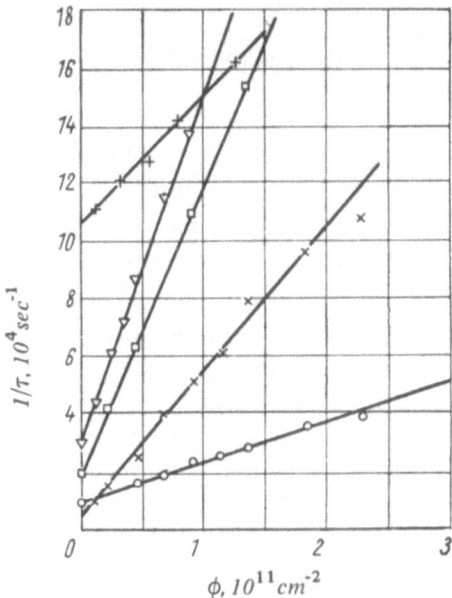

Figure 53. Dependence of $1/\tau$ on the integrated flux of fast neutrons in n-type silicon (○) and in p-type silicon (▽, x, □, +) of various initial carrier concentrations /204/.

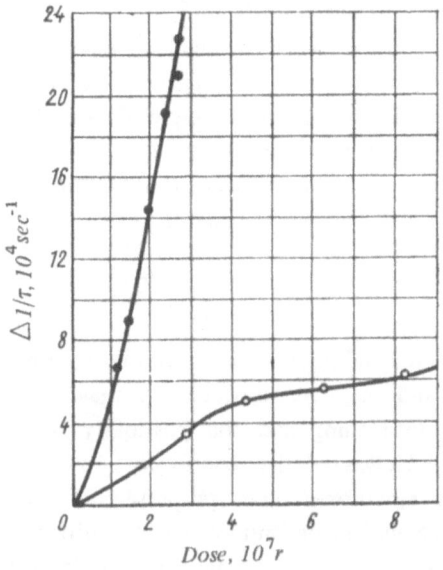

Figure 54. Dependence of the increase in $1/\tau$ on the γ-radiation dose in silicon grown from a melt (○) and in silicon obtained by floating-zone melting. (●)

$$K = [N_A \sigma_d(E) \, \bar{v}(E)] \sum_i^n \eta_i \, (\sigma_{pi} \bar{v}_p f_{ni} + \sigma_{ni} \bar{v}_n f_{pi}). \tag{128}$$

The product in the square brackets of (128) represents the rate of production of structural defects per unit radiation dose. It depends both on the type of radiation and on its energy spectrum. The product to the right of the Σ sign represents the recombination properties and the effectiveness of any center formed as a result of the production of an imperfection in the crystal lattice. In this product η_i denotes the rate of introduction of the i-th center; σ_{pi} and σ_{ni} are the hole and electron capture cross-sections of those centers; v_p and v_n are the mean thermal velocities of a hole and an electron, and f_{pi} and f_{ni} are functions which are determined by the position of the Fermi level with respect to the energy levels of the i-th centers; those functions determine the degree of filling of centers with electrons or holes.

It has been found that despite the fact that irradiation introduces into the forbidden gap a system of levels /213, 214/ the lifetime in an irradiated semiconductor is determined only by one or two dominant centers. Therefore the Shockley-Read recombination statistics are most often used to determine the lifetime and the properties and parameters of the introduced levels /215/.

The K-coefficient can be determined by measuring the lifetime of minority carriers in irradiated materials. This coefficient can also be determined by irradiation of semiconductor devices. For transistors this coefficient can be calculated from data on the rate of change of the reciprocal of the base current transport coefficient and from the transit time of injected carriers through the base:

$$K = 1/t_{tr} \ \frac{d(1/B)}{d\Phi}. \tag{129}$$

The use of semiconductor devices for the determination of K has both positive and negative aspects. Its main advantages include simplicity, remote reading and the possibility to obtain numerous statistical data. However, the coefficients obtained in this way characterize the radiation-induced properties not of the material of which the device was made, but of a rather different material since formation of p-n junctions changes the initial properties of the material.

Differences between the values of K for Ge and Si (and between the dependences of K on the radiation energy) can be attributed to differences in the defect production rates, as a result of differences in atomic weight, formation cross-section and nature of stable (at a given temperature) defects. Data on the number of removed carriers (in low-resistivity materials) may be used to compare the number of defects produced by one fission-spectrum fast neutron.

Cleland /216,217/ found that on the average 3.2 electrons (per neutron/cm^2) are removed in n-type germanium with a resistivity of several ohms/cm. Binder /218/ irradiated the specimens with fission neutrons and obtained higher values (8±1 electrons/neutron/cm^2). Kantz /219/ found that in n-type Si with a resistivity of 1 ohm-cm (obtained by zone melting) the number of removed electrons was 11.5 neutron/cm^2 (with a mean energy close to the mean energy of the fission spectrum (1.95 MeV)), Kantz found that for silicon with a low oxygen content a linear relationship exists between the rate of carrier removal and the mean neutron energy in the spectrum, and showed that the energy spectrum of neutrons must be taken into account in the study of irradiated semiconductors.

Let us now consider in greater detail the characteristic features of germanium and silicon.

Germanium. Messenger *et al.* /220/ were among the first to determine the values of K for germanium transistors irradiated in reactors. For alloyed p-n-p transistors $K=(2.0\pm0.8)\cdot10^{-8}$ (sec·neutron/cm^2)$^{-1}$ while for n-p-n transistors $K=(4.2\pm0.7)\cdot10^{-8}$(sec·neutron/cm^2)$^{-1}$. In both cases the resistivity of the base material was close to 2 ohm·cm. The specimens were irradiated in the pulsed critical setup "Godiva" at the MTR at Los Alamos /221/. Other specimens were irradiated with neutrons produced by a beryllium target bombarded with deuterons (up to 20 MeV). All results were normalized to bombardment with 4 keV neutrons with a spectrum similar to that of the Godiva assembly. The authors devoted most of their attention to alloyed transistors while for diffused transistors with heterogeneous alloy bases they made only approximate determinations of K whose numerical value was one order of magnitude higher than that for alloyed transistors $(2.0\cdot10^{-7}$(sec·neutron/cm^2)$^{-1}$). This difference between the values of K for alloyed and diffused (alloyed-diffused) transistors was later confirmed by Taylor /222/.

In addition to the values of K, Messenger *et al.* found the parameters of recombination levels associated with a change in the lifetime of n- and p-type germanium. Analyzing the dependence of the reciprocal current transport coefficient on the injection levels for various integrated neutron fluxes they found that those levels can be expected to be 0.23 ± 0.02 eV below the bottom of the conduction band and that the hole and electron capture cross sections would be $\sigma_p = 10^{-15} \mathrm{cm}^2$ and $\Phi_n = 4\cdot10^{-15}$ cm^2.

Using a different method Messenger /188/ found that medium and high injection levels produced in germanium transistors of any type are 0.18 eV below the bottom of the conduction band. At low injection levels the influence of levels with an activation energy of 0.24 eV becomes more important.

The hole capture cross section of the $E_c - 0.18$ eV level for low-power alloyed 2N404 transistors is $\sigma_p = 1.1\cdot10^{-15}$ while for n-p-n 2N1308 alloyed transistors $\sigma_p = 2.5\cdot10^{-15} \mathrm{cm}^2$. The electron capture cross section is considerably lower ($\sigma_n = 5.9\cdot10^{-16}$ cm^2 and $\sigma_n = 2.8\cdot10^{-16}$ cm^2 respectively). Hence, Messenger concluded, that the implanted centers are negatively charged with respect to the captured hole while they are neutral with respect to electrons.

If the recombination takes place on centers whose levels, like the Fermi levels, lie in the upper half of the forbidden gap, then K can be written as follows:

$$K = \frac{dN_t}{d\Phi} \bar{v}_p \sigma_p \frac{1 + \Delta n/n_0}{1 + n_1/n_0 + \Delta n/n_0} \ . \tag{130}$$

where $dN_t/d\Phi$ is the rate of implantation of recombination centers; n_0 is the equilibrium electron concentration; $\Delta n/n_0$ is the injection level for the excitation of non-equilibrium carriers; and n_1 is the concentration of electrons in the conduction band if the Fermi level coincides with the E_i-level of recombination centers.

Equation (130) indicates that in strongly doped n-type germanium, in which all recombination centers are filled, the coefficient K should have a maximum value which does not depend on the concentration of carriers:

$$K_{max} = \frac{dN_t}{d\Phi} \bar{v}_p \sigma_p. \tag{131}$$

This maximum value of K can be found in materials with a Fermi level several kT higher than the $E_t = E_c - 0.18$ eV level, i.e. when the carriers concentration exceeds $10^{17} \mathrm{cm}^{-3}$. For transistors with a heterogeneous alloy base and a carrier concentration (at the emitter boundary) exceeding $10^{17}\,cm^{-3}$ the coefficient K is close to the maximum value.

Figure 55 shows experimental data on $1/K$ /187/ which illustrate the experimentally found dependence of K on the concentration of electrons. The almost symmetrical distribution of these values with respect to the intrinsic concentration indicates that it is impossible to state which type of conductivity is preferable from the standpoint of the changes in τ under the influence of radiation. Apparently the radiation stability is almost the same at least for alloyed p-n-p and n-p-n germanium transistors.

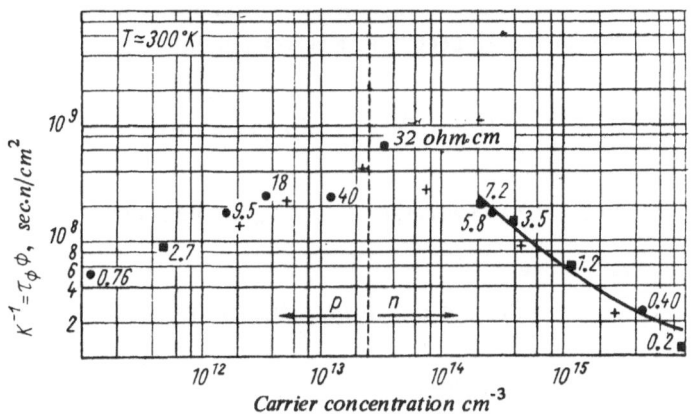

Figure 55. Dependence of $1/K$ on the equilibrium electron concentration in Ge; $+$/205,206/; $\bullet$/223/, $\blacksquare$/189/ (the figures indicate the resistivity, in ohm-cm).

The position of levels in the forbidden gap found by Messenger almost coincides with the position of levels found by other authors who irradiated germanium with neutrons /205, 224, 225/ and with γ-rays from cobalt sources /208, 210/ The only difference concerns the activation energy of levels in the lower half of the forbidden gap. Sharendo and Smirnov /226/ found for those levels a value of $E_t = E_v + 0.26\,eV$. Curtis and Crawford /73/ believe that these levels lie $0.36\,eV$ or $0.28\,eV$ above the edge of the valence band depending on the type of dopant (Sb or As respectively). In specimens with levels in the lower half of the forbidden gap the lifetime increased with the increase in the excitation level. Curtis /227/ who analyzed the conflicting results on the irradiation of germanium (which reported different levels) assumed that all levels, i.e. $E_{ta} = E_v + (0.24 - 0.36)$ eV and $E_{tb} = E_c - (0.18-0.2)$ eV are active. However their activities differ and depend on the injection or excitation level. At very low levels ($10^{-4} \leqslant \Delta n/n_0 \leqslant 10^{-3}$), at which τ is determined from the decrease in photoconductivity, the levels in the lower half of the forbidden band are predominant. The lifetime associated with such levels increases with increasing $\Delta n/n_0$ ratio. On the other hand, the lifetime associated with the $E_c - 0.2$ eV levels continuously decreases with increasing $\Delta n/n_0$ and at a certain value of $\Delta n/n_0$ it becomes the determining factor.

According to the Shockley-Read recombination statistics the lifetime associated with deep-lying levels can be found from:

$$\tau_a = (p_{1a} + \Delta n)/C_{na} N_{ta} (n_0 + \Delta n). \tag{132}$$

For the $E_c - 0.2$ eV levels the value of τ_b can be found from:

$$\tau_b = (n_0 + n_{1b} + \Delta n)/C_{pb} N_{tb} (n_0 + \Delta n), \tag{133}$$

where $C_n = v_n \sigma_n$ and $C_p = v_p \sigma_p$.

Hence, the lifetime would be equal to:

$$\tau = (1/\tau_a + 1/\tau_b)^{-1}$$

or

$$\tau = \left[\frac{C_{na} N_{ta} (n + \Delta n)}{p_{1a} + \Delta n} + \frac{C_{pb} N_{tb} (n_0 + \Delta n)}{n_0 + n_{1b} + \Delta n} \right]^{-1}. \tag{134}$$

The function $\tau = F (\Delta n/n_0)$ on a relative scale for $n_0 = 5 \cdot 10^{14}$ cm^{-3}, $P_{ia} = 10^{13}$ cm^{-3} and for two values of n_{ib} ($2 \cdot 10^{15}$ and $4 \cdot 10^{15}$ cm^{-3}) is given in Figure 56. This figure shows also $1/\tau_a$ and $1/\tau_b$ as a function of $\Delta n/n_0$ and indicates that for $\Delta n/n_0 \geqslant 8\%$ the lifetime is determined by the $E_c - 0.2$ eV levels and decreases with increasing $\Delta n/n_0$ ratio.

Irradiation of germanium with electrons or γ-rays produces $E_c - 0.2$ eV ($E_c - 0.22$ eV according to the latest data of Spitsin /228/) levels in the forbidden gap. However, the hole capture cross-section of those levels is

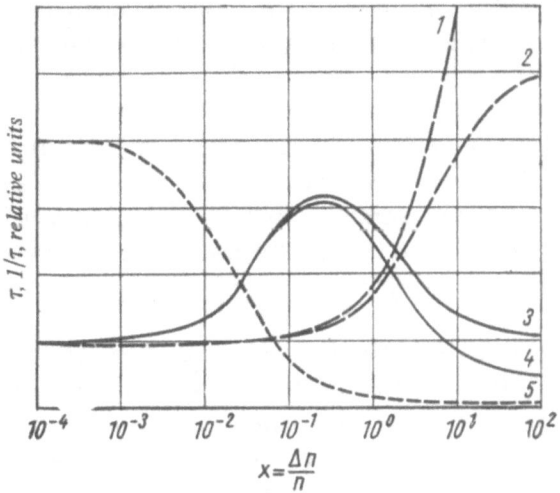

Figure 56. Dependence of the lifetime of minority carriers in germanium on the injection level in the presence of two active recombination levels, when $n_0 = 5 \cdot 10^{14}$ cm^{-3}, $p_{1a} = 10^{13}$ cm^{-3} and $1/\tau_a = 4/\tau_b$

1 - $1/\tau_b$ ($n_{1b} = 4 \cdot 10^{15}$ cm^{-2}); 2 - $1/\tau_b$ ($n_{1b} = 2 \cdot 10^{15}$ cm^{-3});
3 - $\tau(n_{1b} = 2 \cdot 10^{15}$ cm^{-3}); 4 - $\tau(n_{1b} = 4 \cdot 10^{15}$ cm^{-3}); 5 - $1/\tau_a$;

small. Its most probable value is about $4 \cdot 10^{-16}$ cm^2 /206/,which is nearly one order of magnitude less than the cross section of similar centers produced by neutron irradiation.

Bertolotti and Sette /229/ believe that this difference is due not to differences in the nature of defects produced by γ-rays and neutrons but to additional recombination of the disordered regions (defect clusters) which are produced by neutron irradiation. If we assume that the recombinations of defects are not influenced by those of clusters we can calculate the lifetime of electrons from the following equation:

$$\frac{1}{\tau} = C_p N_t \left(\frac{n_0}{n_0 + n_1} \right) + \frac{C_{rp} N_r n_0}{n_0 + (C_{rp}/C_{rn}) p_0'} . \tag{135}$$

The first term represents the recombination of simple defects while the second term represents the recombination of clusters (N_t -concentration of simple defects, N_r -concentration of clusters and of disordered regions). The capture probabilities of holes C_{rp} and of electrons C_m by defect clusters are equal to:

$$C_{rn} = f \pi r_{eff}^2 \overline{V}_p$$

$$C_{rp} = f \pi r_{eff}^2 \overline{V}_n$$

153

respectively, where r_{eff} is the effective radius of (arbitrarily) spherical clusters; f is a coefficient which represents the effectiveness of defect clusters as recombination centers, relative to simple defects. In equation (135) p_0' stands for the equilibrium concentration of holes in clusters. According to Bertolotti and Sette $p_0'=2.2\cdot10^{17}\,\mathrm{cm}^{-3}$. It has been found /229/ that most of the results obtained by Cleland and Curtis, who irradiated germanium with γ-rays from a cobalt source, with reactor neutrons and with 14 MeV neutrons, can be satisfactorily correlated by taking $f=9\cdot10^{-2}$.

Silicon.

The early data of Messenger et al. /220/ ($K=(3.9\pm1.1)\cdot10^{-7}$ and $(3.6\pm1.2)10^{-7}$ (sec$\cdot$n/cm^2)$^{-1}$ for n- and p-type Si respectively) indicate that both p-type and n-type silicon have a poorer radiation stability than germanium of similar resistivity (2 ohm$\cdot$cm) /220/. It was also found /204/ that the coefficient K for silicon used in semiconductor devices, is not greatly influenced by the concentration of majority carriers or by the injection levels of minority carriers, and it changes only by a factor of 3–4 if the conductivity of the material and the injection level are changed by 3–4 orders of magnitude.

In an investigation of the temperature effect on the reciprocal of the base current transport coefficient, Messenger found that in n-type silicon the dominating recombination centers are those with E_c-0.18 eV levels (in p-type silicon those with E_v +0.3 (0.27) eV levels). He calculated the carrier capture cross-sections: $\sigma_p=5.5\cdot10^{-13}\,\mathrm{cm}^{-3}$; $\sigma_n=7.0\cdot10^{-14}\,\mathrm{cm}^2$ for E_c -0.18 eV and $\sigma_p=6.4\cdot10^{-15}\,\mathrm{cm}^2$, and $\sigma_n=2.9\cdot10^{-15}\,\mathrm{cm}^2$ for E_v +0.30eV.

The same levels were found by many authors after irradiation with neutrons, electrons and γ-rays, in silicon grown by the Czochralski method and containing a large number of atoms of dissolved oxygen (about $10^{18}\,\mathrm{cm}^{-3}$) /232, 233, 236/. Data on the carrier capture cross-sections of those levels are given in Table 9.

In γ-irradiated phosphorus-doped "oxygen-free" silicon ($<10^{16}$ atoms of oxygen/cm^3) a dominant role is exerted by E centers which consist of associations of vacancies with phosphorus atoms, with E_c^-0.4 eV levels /50/. Some data on the carrier capture cross section of E centers are given in Table 9. The rate of production of those centers is a function of the relative content of phosphorus and oxygen /238/.

The E_c-0.4 eV level, corresponding to E centers, has also been found in silicon irradiated with neutrons, electrons and γ-rays; however,

Table 9. Properties of dominant recombination centres in irradiated silicon.

Material and its characteristics	Radiation	Experimental method	Level position, eV	Rate of introduction cm^{-1}	Hole capture cross-section 10^{-14} cm^{-2}	Electron capture cross-section 10^{-16} cm^2	Literature
FZ150 (p)	Co^{60}	$1/\tau = F(\Phi)$ $\tau = F(1/T)$	$E_c - 0.17$	$1.3 \cdot 10^{-3}$	1.5	1.8	/230/
CZ85 (p)		Ditto	$E_c - 0.17$	$1.3 \cdot 10^{-3}$	1.4	2.4	/230/
CZ40 (n)			$E_c - 0.17$	$1.3 \cdot 10^{-3}$	1.5	—	/230/
FZ50 (n)			$E_c - 0.40$	$1.5 \cdot 10^{-4}$	1.1	—	/230/
FZ250 (n)			$E_c - 0.40$	$\ll 10^{-4}$	1.1	—	/230/
FZ70 (n)			$E_c - 0.40$	$1.2 \cdot 10^{-4}$	4.0	—	/230/
FZ77 (n)			$E_c - 0.17$	$1.2 \cdot 10^{-4}$	1.0	—	/231/
FZ77* (n)			$E_c - 0.40$	$1.4 \cdot 10^{-4}$	9.0	—	/231/
FZ32 (n)			$E_c - 0.17$	$1.4 \cdot 10^{-4}$	1.0	—	/231/
FZ32 (n)			$E_c - 0.40$	$3.4 \cdot 10^{-4}$	9.0	—	/231/
CZ7 (n)	Electrons (0,7 MeV)	$\tau = F(1/T)$ $R_{Hall} = F(1/T)$	$E_c - 0.16$	0.18	—	—	/232/
CZ7 (n)			$E_v + 0.31$	0.18	2.8	1	/232/
CZ5 (p)			$E_v + 0.29$	0.005	—	—	/232/
CZ5 (p)	Reactor neutrons	$R_{Hall} = F(1/T)$	$E_c - 0.16$	0.005	—	—	/233/
CZ5 (p)			$E_v + 0.3$ (± 0.01)	0.35	—	9.5	/233/
CZ5 (p)			$E_v + 0.16$	0.35	—	—	/233/

Table 9 cont.

Material and its characteristics	Radiation	Experimental method	Level position, eV	Rate of introduction cm^{-1}	Hole capture cross-section 10^{-14} cm^2	Electron capture cross-section 10^{-16} cm^2	Literature
CZ (n)	Co^{60}	Photoconductivity spectrum	$E_c - 0.16$	0.35	2.5	—	/234/
CZ (n)	Electrons (Sr^{90} — Y^{90})	$R_{Hall} = F(1/T)$	$E_c - 0.16$	0.35	4.0	1.0	/235/
FZ300 (n)	Co^{60}		$E_c - 0.16$	$9.4 \cdot 10^{-4}$	—	—	/236/
FZ1500 (n)			$E_c - 0.16$	$7.5 \cdot 10^{-4}$	—	—	/236/
FZ1000 (n)			$E_v + 0.27$	$7.5 \cdot 10^{-4}$	—	—	/236/
FZ10000 (p)			$E_v + 0.27$	$3.4 \cdot 10^{-4}$	—	—	/236/
CZ —	Co^{60}	$R_{Hall} = F(1/T)$	$E_c - 0.16$	0.001	—	—	/237/
FZ (0.09—13.7) (n)			$E_c - 0.17$	$(0.5-1.5) \cdot 10^{-3}$ to $1.5 \cdot 10^{-5}$	—	—	/237/
FZ (0.00—13.7) (n)			$E_c - 0.40$	$3.5 \cdot 10^{-3}$ depending on P concentration	—	—	/237/ /238/
CZ7 (p)			$E_v + 0.35$	10^{-4}	—	—	/239/
CZ16 (p)			$E_v + 0.35$	$0.85 \cdot 10^{-4}$	—	—	/239/
CZ180 (p)			$E_v + 0.35$	10^{-4}	—	—	/239/
FZ5 (p)			$E_v + 0.21$	$5 \cdot 10^{-5}$	—	—	/239/
FZ15 (p)			$E_v + 0.21$	$2 \cdot 10^{-5}$	—	—	/239/
FZ50 (p)			$E_v + 0.28$	$\ll 10^{-6}$	—	—	/239/
FZ110 (p)	Reactor neutrons	$R_{Hall} = F(1/T)$ and photoconductivity spectrum	$E_c - 0.16$	?	—	$\sim 10^{-5}$	/51/
PZ110 (p)			$E_v + 0.45$	?	—	$\lesssim 10^{-3}$	/51/
FZ110 (p)			$E_v + 0.38$	0.8	—	$5 \cdot 10^{-3}$	/51/
FZ110 (p)			$E_v + 0.30$	0.08	—	3	/51/
C (1—12) (n)	Electrons (1 MeV)		$E_v + 0.3$ (±0.01)	—	—	—	/51/

156

Table 9 cont.

Material and its characteristics	Radiation	Experimental method	Level position, eV	Rate of introduction cm⁻¹	Hole capture cross-section 10^{-14} cm²	Electron capture cross-section 10^{-16} cm²	Literature
C (1—12) (p)	Electrons (1 MeV		$E_v + 0.27$ (± 0.01)	$5.2 \cdot 10^{-2}$	—	—	/51/
			$E_v + 0.25$ (± 0.01)	—	—	—	/51/
			$E_v + 0.21$ (± 0.01)	$3.4 \cdot 10^{-2}$	—	—	/51/
			$E_v + 0.19$ (± 0.01)	$7.3 \cdot 10^{-3}$	—	—	/51/
? (n)	Reactor neutrons	$B = F(1/T)$ in transistors	$E_c - 0.18$	?	$5.5 \cdot 10$	$7 \cdot 10^{-14}$	/188/
? (p)			$E_v + 0.30$	?	0.6	$2.9 \cdot 10^{-16}$	/188/

Note: The notation in the table is as follows: FZ—silicon obtained by floating-zone melting in vacuum; CZ—silicon produced by the Czochralski method and containing $10^{17} - 10^{18}$ cm⁻³ oxygen; the numbers designate the resistivity (ohm.cm) and the letters in parentheses show the type of conductivity.

data on capture cross-sections are given only for irradiation with γ-rays and electrons, since no recombination centers are produced by neutron irradiation. Stein /241/ determined the hole capture cross-section of recombination centers formed in n-type silicon by neutron and γ-irradiation ($\sigma_p = 1.9 \cdot 10^{-14}$ cm^2 for γ-irradiation and $\sigma_p = 7.7 \cdot 10^{-14}$ cm^2 for neutron irradiation). The author did not determine the level position but since he used oxygen-free silicon, in both cases it may be assumed that those cross-sections are related to E centers.

The data in Table 9 show that the same is true for the carrier capture cross-section of A centers (E_c–0.17 eV): the cross sections of centers, formed by neutron irradiation, are always greater than the cross-sections of the same centers formed by γ-radiation. This fact indicates that in neutron irradiated silicon recombination takes place almost exclusively in defect clusters and not on simple centers, which are also produced by neutron irradiation and whose nature can be studied by determining the dependence of their lifetime on $1/T$. A similar assumption has been made by Bertolotti and Sette for germanium. The above approach was used by Stein to explain the differences in defect annealing and in the temperature dependences of the defect production rates by neutrons and γ-radiation. Defects produced by neutron bombardment disappear during annealing at a rate which increases monotonically with increasing temperature while defects produced by γ-radiation are annealed at a rate which becomes noticeable and then rapidly increases only after heating to a certain "threshold" temperature.

The rate of defect production in Stein's experiments, or more exactly, the coefficient K for neutron irradiation is not influenced by temperature in the range $70-300^\circ$K and is equal to $5.8 \cdot 10^{-6}$ (neutron/cm^2)$^{-1}$ (sec)$^{-1}$ for > 10 keV fission spectrum neutrons. For γ-radiation from a Co^{60} source the coefficient K is $2.8 \cdot 10^{-10}$ (γ-quanta/cm^2)$^{-1}$ (sec)$^{-1}$ at 300°K and decreases with decreasing temperature. The silicon specimens used in the experiments were phosphorus-doped and had a resistivity of 6 ohm·cm.

The above assumption is confirmed by the data of Curtis /242/ who found that the carrier lifetime in neutron-irradiated silicon is not influenced by the dopant or by the concentration of oxygen, but only by the type of conductivity and by the concentration of majority carriers.

In order to explain the results of Curtis, Messenger /243/ suggested a two-level model which represents all active recombination levels by two discrete levels, one of which is in the upper half of the forbidden gap and the second one in the lower half. It has been found that the parameters of these levels are not influenced by the type of dopant or by the concentration of oxygen and therefore they can be expected to be equally active in n- and p- materials. Since those parameters are not interrelated, the total lifetime can be determined by summing the reciprocals of the "partial" lifetimes associated with each level:

$$\frac{1}{\tau} = \frac{1}{\tau_1} + \frac{1}{\tau_2} = K\Phi. \tag{136}$$

According to the Shockley-Read recombination statistics the coefficient K can be calculated from:

$$K = \frac{1}{\frac{1}{C_{p1}R_1}\left(\frac{n_0+n_1+\Delta n}{n_0+p_0+\Delta n}\right) + \frac{1}{C_{n1}R_1}\left(\frac{p_0+p_1+\Delta n}{n_0+p_0+\Delta n}\right)} +$$
$$+ \frac{1}{\frac{1}{C_{p2}R_2}\left(\frac{n_0+n_1+\Delta n}{n_0+p_0+\Delta n}\right) + \frac{1}{C_{n2}R_2}\left(\frac{p_0+p_1+\Delta n}{n_0+p_0+\Delta n}\right)}. \tag{137}$$

Here, as above, n_0 and p_0 are the equilibrium concentrations of electrons and holes; n_1, p_1, n_2, and p_2 are the concentrations of electrons and holes if the Fermi levels coincide with the positions of the recombination levels, in this case with level 1 and with level 2; Δn is the non-equilibrium carrier concentration and $C_{p1}R_1, C_{n1}R_1, C_{p2}R_2$ and $C_{n2}R_2$ are the products of the rate of introduction of levels and of the probability of carrier capture by these levels.

For small injection levels, corresponding to the experimental data of Curtis, equation (137) can be written in a simpler form:

$$K_n = \frac{C_{p1}R_1}{1+n_1/n_0} + \frac{C_{p2}R_2}{1+\frac{C_{p2}}{C_{n2}}\cdot\frac{p_2}{n_0}} \tag{138}$$

for n-type semiconductors and

$$K_p = \frac{C_{n1}R_1}{1+\frac{C_{n1}}{C_{p1}}\cdot\frac{n_1}{p_0}} + \frac{C_{n2}R_2}{1+p_2/p_0} \tag{139}$$

for p-type semiconductors.

The statistical data of Curtis were used by Messenger to calculate the positions of those two "average" levels in the forbidden gap and of other characteristics which are included in the equations for K_n and K_p; the upper level was 0.265 eV below the bottom of the conduction band while the lower level was 0.31 eV above the edge of the valence band. The position of the lower level agrees with that of the level found in the irradiation of p-type silicon. Other parameters had the following values:

$$C_{p_1}R_1 = 0.37 \cdot 10^{-6} \ (\text{sec.neutron/cm}^2)^{-1};$$
$$C_{n_1}R_1 = 0.40 \cdot 10^{-5} \ (\text{sec.neutron/cm}^2)^{-1};$$
$$C_{n_2}R_2 = 0.68 \cdot 10^{-5} \ (\text{sec.neutron/cm}^2)^{-1};$$
$$C_{p_2}R_2 = 0.76 \cdot 10^{-6} \ (\text{sec.neutron/cm}^2)^{-1};$$
$$n_1 = 2.0 \cdot 10^{14} \ \text{cm}^{-3} \ \text{and} \ p_2 = 1.3 \cdot 10^{13} \ \text{cm}^{-3}.$$

If we substitute those values in equations (138) and (139) we obtain the following expressions for the dependence of K_n and K_p on the resistivity:

$$K_p = \frac{1{,}4 + 8{,}6 \cdot 10^{-2}\rho + 1{,}2 \cdot 10^{-3}\rho^2}{1 + 3{,}8 \cdot 10^{-2}\rho} \tag{140}$$

and

$$K_n = -\frac{2{,}1 + 0{,}18\rho + 9{,}0 \cdot 10^{-5}\rho^2}{1 + 1{,}4 \cdot 10^{-2}\rho} \tag{141}$$

The experimental data of Curtis and the corresponding functions (equations (140) and (141)) are shown in Fig. 57.

The level constants determined by Messenger may be used to plot the dependence of the reciprocal of K on the resistivity and injection level $x = \Delta n/n_0 (\Delta n/p_0)$ for n- and p-type silicon. These functions are shown in Fig. 58.

As the resistivity decreases, the value of $1/K$ asymptotically approaches a limit that increases with increasing injection level. The changes in its absolute value are the greatest at low injection levels. At very high injection levels its value depends on the resistivity and is close to 10^6 sec. n/cm^2. For a low-resistivity material the asymptotic values of K (at low injection levels) for n- and p-type Si are $C_{p_1}R_1 + C_{p_2}R_2$ and $C_{n_1}R_1 + C_{n_2}R_2$ respectively, indicating that n-type Si is more resistant to neutrons.

160

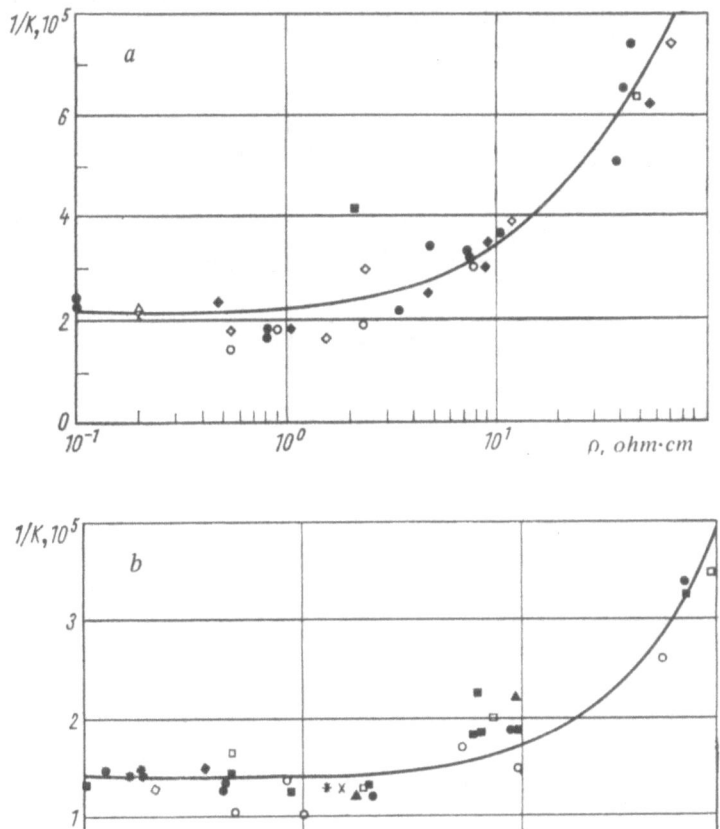

Figure 57. Dependence of the values of $1/K$ (for silicon) on the resistivity during neutron irradiation of samples with p- (*a*) and n-type (*b*) conductivity (several specimens were studied).

The temperature dependence of the coefficient K is also calculated using equation (137). The final expressions for $K=F(T)$, normalized to room temperature T_1, can be written as follows:

$$K_n = \frac{C_{p1}R_1}{\dfrac{C_{p1}}{C_{n1}} \left(\dfrac{\Delta n}{n_0 + \Delta n}\right) + 1 + \left(\dfrac{n_0}{n_0 + \Delta n}\right) \times \left(\dfrac{n_1}{n_0}\right) D_1} +$$

161

$$+ \cfrac{C_{n2}R_2}{\cfrac{C_{n2}}{C_{p2}}+\cfrac{\Delta n}{n_0+\Delta n}+\left(\cfrac{n_0}{n_0+\Delta n}\right)\times\left(\cfrac{p_2}{n_0}\right)D_2} \; ; \qquad (142)$$

$$K_p = \cfrac{C_{p1}R_1}{\cfrac{C_{p1}}{C_{n1}}+\cfrac{\Delta n}{p_0+\Delta n}+\left(\cfrac{p_0}{p_0+\Delta n}\right)\times\left(\cfrac{n_1}{p_0}\right)D_1} +$$

$$+ \cfrac{C_{n2}R_2}{\cfrac{C_{n2}}{C_{p2}}\left(\cfrac{\Delta n}{p_0+\Delta n}\right)+1+\left(\cfrac{p_0}{p_0+\Delta n}\right)\times\left(\cfrac{p_2}{p_0}\right)D_2} \; . \qquad (143)$$

$$D_1 = T_1\left(\frac{T}{T_1}\right)^{3/2}\exp\left[38.6\left(1-\frac{T_1}{T}\right)\Delta E_1\right] ;$$

$$D_2 = T_1\left(\frac{T}{T_1}\right)^{3/2}\exp\left[38.6\left(1-\frac{T_1}{T}\right)\Delta E_2\right] .$$

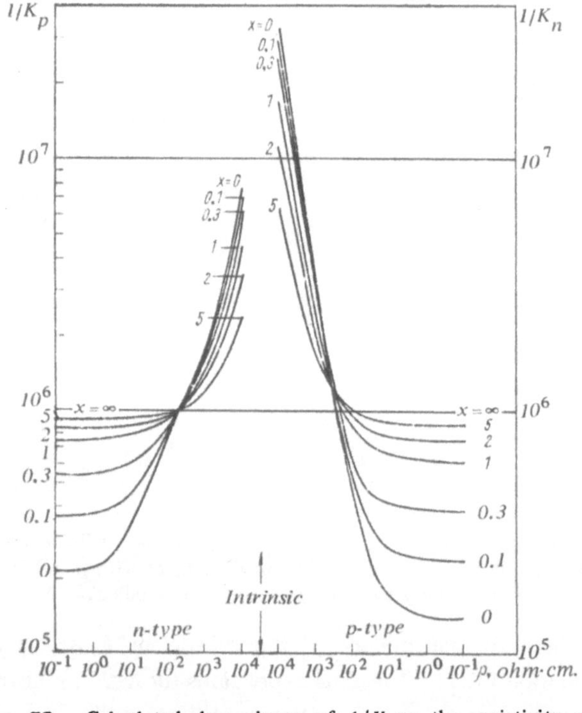

Figure 58. Calculated dependence of $1/K$ on the resistivity of neutron irradiated silicon and on the injection level of non-equilibrium carriers. These calculations were based on the relationship between the equilibrium carrier concentration and the resistivity of the material: $p_n = 5\cdot10^{15}/n_0$ and $p_p = 2.5\cdot10^{16}/n_0$.

The dependence of $1/K$ on the reciprocal temperature at various resistivities and injection levels is shown in Figure 59. If the injection level increases, K decreases at low temperatures but increases at high temperatures. The limits of the temperature variation of K depend on the resistivity. The value of K is greatly influenced by the temperature if the resistivity is high but the influence is small in the range of $250-500°K$ if the resistivity of the material is low.

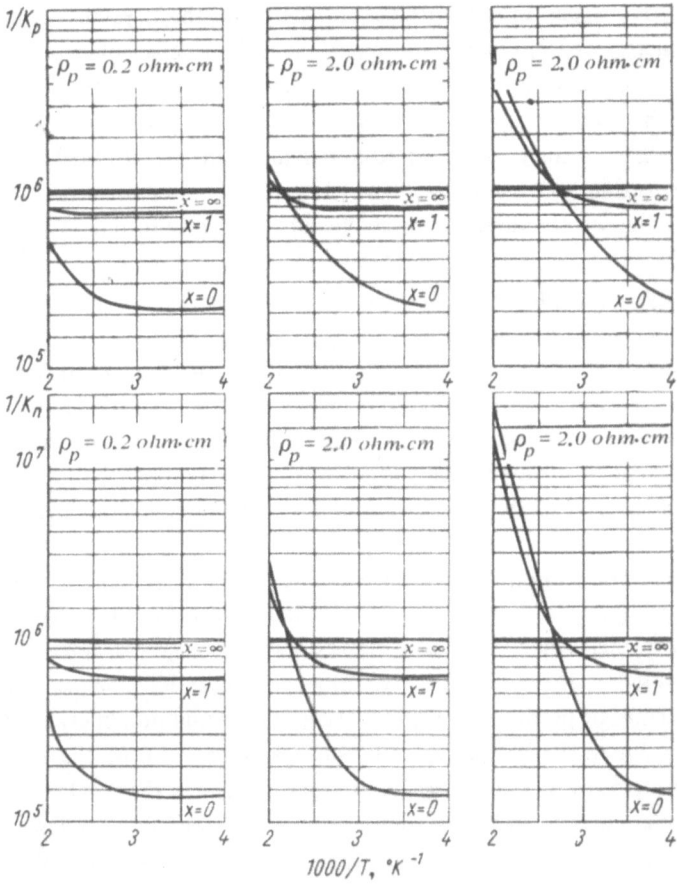

Figure 59. Temperature dependence of $1/K$ for neutron-irradiated silicon of various resistivities and at various injection levels of non-equilibrium carriers.

The two-level model for silicon, suggested by Messenger, was used for an analysis of the experimental temperature dependence of lifetime

163

for low injection levels and for an analysis of the dependence of the base current transport coefficient of *p-n-p* and *n-p-n* transistors on the temperature and the injection level. The experimental results agreed satisfactorily with data predicted on the basis of the above model.

6. Surface Effects in Irradiated Transistors

It has been found that the degradation of transistor parameters during irradiation is due both to structural damage in the crystal lattice and to changes in the surface properties of the crystals. The main parameters of transistors and, in particular, the reverse collector current I_{co}, are very sensitive to the surface state. A change in the recombination properties of surface layers, particularly in the immediate vicinity of the emitter *p-n* junction, affects first the base current transport coefficient, which can either increase or decrease depending on the direction of change of the surface recombination rate *s*. The production of inversion layers and of surface channels in the vicinity of *p-n* junctions leads to a considerable increase in the reverse currents across the junctions.

Surface processes excited by ionizing radiation occur even at small radiation doses which, usually, are insufficient to produce appreciable bulk damage. The changes in the base current transport coefficient, associated with surface processes, approach saturation but this does not apply to changes in the reverse collector current (of similar origin).

Figure 60 shows a typical dependence of the reciprocal of the base current transport coefficient on γ-radiation dose for high-frequency *n-p-n* planar silicon 2N1613 transistors which were irradiated in a passive state. In the case of large doses, the linear dependence is due to bulk damage produced by γ-radiation. The initial non-linear relationship is due to changes produced by surface processes. Both the bulk and surface effects, and also the factors that determine the total change in the base current transport coefficient can be regarded as independent of each other. Therefore, if we deduct from the overall effect the bulk effects (Figure 60) we obtain the dependence of the surface component (of the total effect) on the radiation dose. This component grows rapidly and approaches saturation at a certain value. Brown /244/ used an exponential function (Fig. 60) to represent the above effects.

It has been found /245/ that the surface effects are caused primarily by ionization phenomena in the oxide layers of crystals and in their

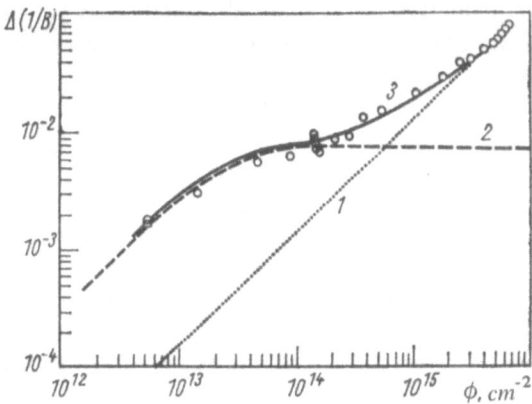

Figure 60. Dependence of the increase in $1/B$ on the integrated 0.53 MeV electron flux in n-p-n planar 2N1613 silicon transistors containing both surface and bulk defects. For bulk defects (1) $\Delta(B^{-1}) = 1.56 \cdot 10^{-17}\Phi$; for surface defects (2) $\Delta(B^{-1}) = 0.008\%[1-\exp(4 \cdot 10^{14}\Phi)]$; experimental line (3).

immediate vicinity. This explains why such effects are so easily induced by high energy radiation as well as by low-energy radiation which cannot cause bulk damage. Figure 61 shows the results of irradiation of

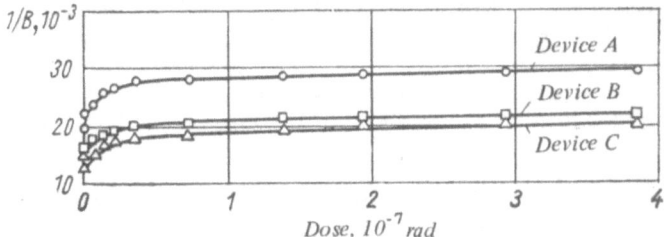

Figure 61. Typical dependence of $1/B$ on the radiation dose. The value of $1/B$ is associated with surface processes taking place under the effect of radiation.

2N1613 transistors with 150 keV X-rays, which cannot, at the doses used in the experiments, produce bulk damage that would lead to such changes in the base current transport coefficient. The results also confirm that it is possible to differentiate between surface and bulk effects in transistors using as a criterion the dependence of those effects on the radiation dose.

It has been found /246/ that the saturation effect and the time needed for- saturation at a constant radiation dose depend on the operating conditions of the transistor under irradiation. A comparison of data obtained under normal operating conditions at various collector

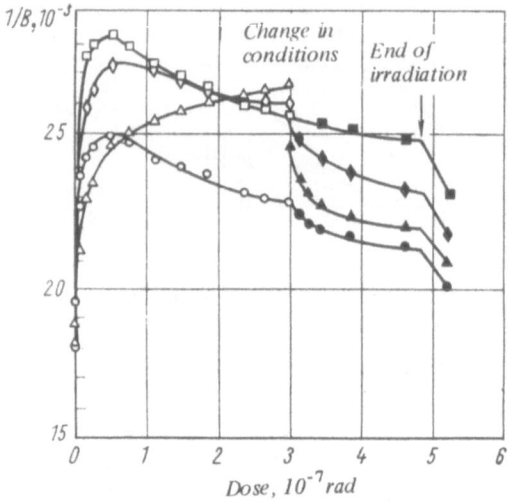

Figure 62. Influence of the operating conditions of a transistor on the changes in base current transport coefficient B caused by irradiation:

$\triangle$ — passive state; $\blacksquare$, $\blacktriangle$, $\square$, $\bullet$ — U_c = 6 V, I_c = 10mA, $\circ$, $\diamond$ U_c = 6V, I_c = 5mA; $\blacklozenge$ U_c = 6V, I_c = 20mA.

currents (Figure 62) or with passive transistors shows that: 1) in active transistors, unlike in passive transistors, the base current transport coefficient will increase rapidly at the beginning, but will subsequently decrease smoothly approaching a value different from zero; 2) all other conditions being equal, this limiting value changes less at higher collector currents.

The initial rapid changes in base current transport coefficient are due to the reverse voltage applied to the collector junction while the subsequent partial recovery is caused by a bias on the emitter junction /246/. Thus, the smallest change in the gain factor of transistors, associated with the surface effect of radiation, would be observed when the transistor is irradiated in an active state, at the maximum possible current and lowest possible voltage on the collector junction.

The above example illustrates only the most obvious cases of surface processes that affect the amplification parameters of transistors. In general, such processes are exhibited in different ways. Thus, for instance, the gain parameters of germanium *p-n-p* transistors are in many cases improved during the initial stage of irradiation, rather than being impaired. Such ambiguity is caused by the fact that both the nature and the intensity of the surface effects, produced by ionizing radiation, are determined by many factors, usually of technological nature. Devices of a given type but produced by different plants may also behave differently. This makes it difficult to predict the extent of gain deterioration under the influence of small radiation doses when the surface effects prevail over bulk effects.

Investigations of surface radiation effects in transistors have been based on the study of the reverse collector current I_{co} which is very sensitive to the surface state. Figure 63 shows the influence of

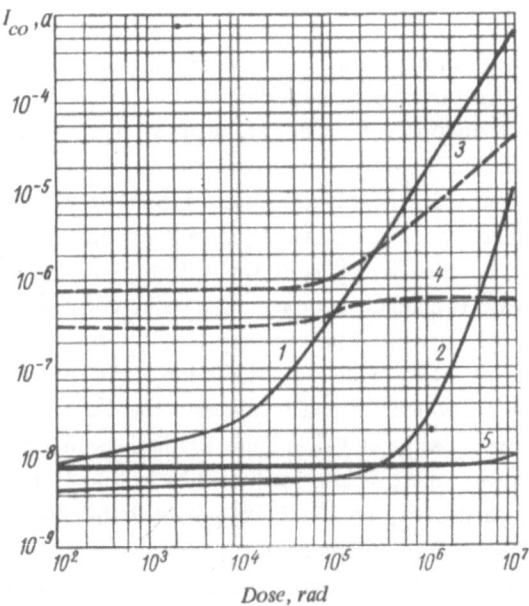

Dose, rad

Figure 63. Dependence of the reverse collector current I_{co} on the dose of γ-radiation for several types of transistors:

1 and 2—diffused *n-p-n* silicon mesa transistors (not evacuated); 3—diffused *p-n-p* germanium transistors (not evacuated); 4—diffused epitaxial *p-n-p* germanium transistor (not evacuated); 5—diffused *n-p-n* silicon-mesa transistor at zero volts on the collector (or with a voltage on collector, evacuated);

167

γ-radiation on I_{co} of transistors manufactured by different methods. The two continuous lines 1 and 2 are related to gas-encapsulated diffused *n-p-n* silicon mesa transistors of the same type.

In the absence of radiation such transistors are very stable and reliable, and exhibit good radiation stability with respect to damage in the bulk. However, they are very sensitive to γ-radiation in the case of a reverse-bias on the collector junction, because of surface processes which take place under the influence of radiation. Figure 63 shows a wide range of doses, corresponding to the beginning of a rapid increase in I_{co} for such transistors and also a considerable scattering of I_{co} values for a given radiation dose. The irradiation of such transistors in the absence of any bias on the collector junction (or of evacuated transistors) caused a slight increase in I_{co} (curve 3) up to a dose of 10^7 rad. The influence of γ-radiation on two 300 MHz gas encapsulated diffused *p-n-p* germanium transistors is shown by dash lines. It was found that the I_{co} of germanium transistors responds more uniformly to γ-radiation /24 /.

The results of an experimental investigation of the reverse collector current for transistors of other types are given in Figure 64 /245/.

After the irradiation is stopped, I_{co} begins to regain slowly its initial value. This process is governed by a law similar to that represented by curve A in Figure 65 /274/. If no voltage is applied during this process, the rate of recovery is much faster (curve B on the same Figure). If however, a reverse bias is applied to the collector junction then I_{co} will start to increase (Figure C) approaching a value corresponding to the extrapolation of curve A. The recovery of I_{co} can be stimulated by numerous factors, one of which is γ-radiation in the absence of bias on the transistor. Sections D, E and F represent the decrease in I_{co} in the course of three subsequent half-minute irradiations in the absence of bias. The recovery of I_{co} is also facilitated by heating to 150°C and by illumination, particularly by UV light, provided that the transistor is not enclosed in a non-transparent can.

The "memory" of germanium and silicon transistors is of great interest. This "memory" is based on the fact that upon repeated irradiation of a transistor (in which I_{co} was restored after it had been increased as a result of irradiation) the reverse current rapidly increases to the value reached at the end of the "primary" irradiation. It is as if the transistor "remembers" the current I_{co} before the recovery.

Further irradiation leads to an increase in I_{co}, which depends on the

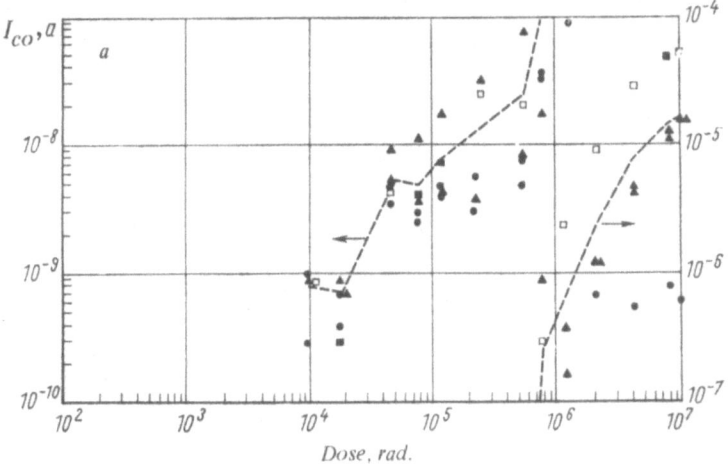

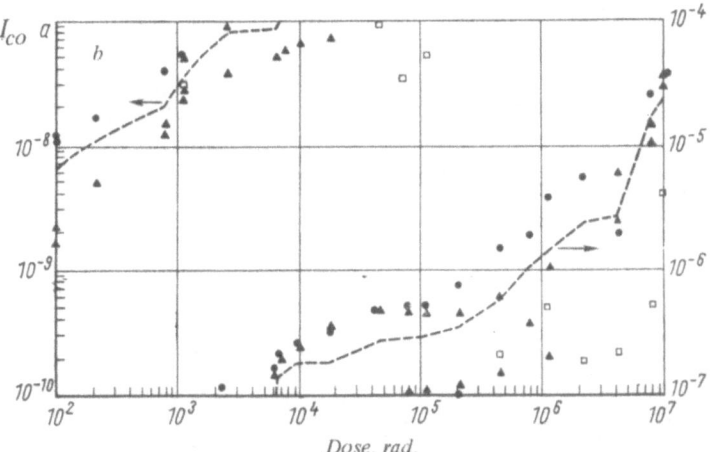

Figure 64. Dependence of the reverse collector current at $U_c =$ 10 V on the γ-radiation dose for p-n-p silicon "mesa" transistors (2N1132) (*a*) and for alloyed silicon p-n-p 2N1676 transistors (*b*). Measurements were carried out on several specimens.

duration of the first irradiation. An example of such memory is shown in Figure 66 /247/. The memory is not lost if the recovery, complete or partial, was the result of γ-irradiation without bias. Heating for several hours at 100°C obliterates almost completely the memory. If the

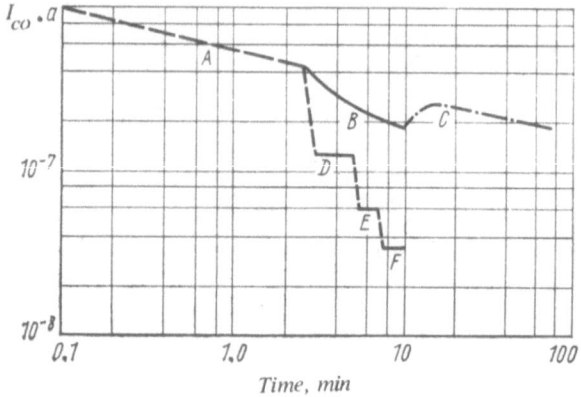

Figure 65. Influence of various conditions on the nature and rate of recovery of the reverse collector current: A—with collector bias; B—after removing the bias; C—reapplied bias; D, E and F, after γ-irradiation for 30 sec. without bias on the collector.

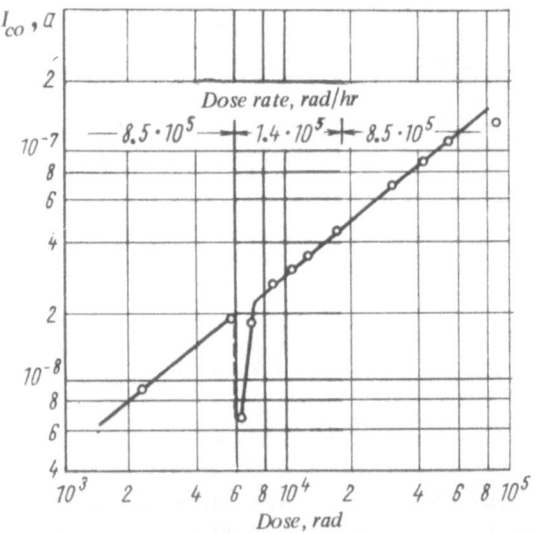

Figure 66. "Memory" effect during γ-irradiation of transistors.

damage to the transistor is not great, the sensitivity to changes in the reverse current during irradiation, which follow such a heat treatment, is much smaller than the sensitivity in the case of the first irradiation.

Most phenomena associated with radiation processes in transistors can be satisfactorily explained using simple models (Figures 67 /247/, 68 /246/). The model in Figure 67 corresponds to a gas encapsulated

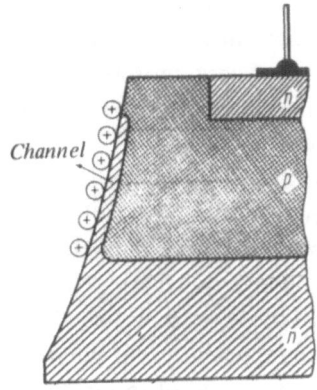

Figure 67. A model representing the formation of inversion layers and channels on the unprotected surface of a base layer.

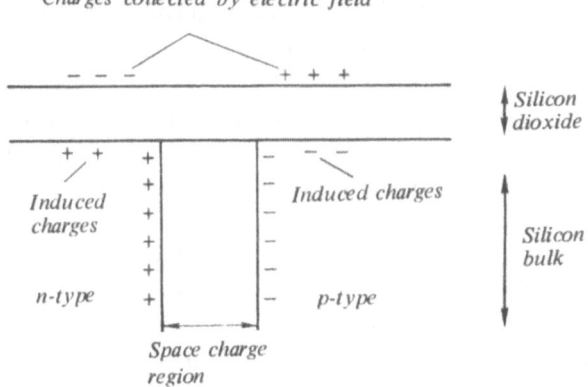

Figure 68. Model of formation of inversion layers in the vicinity of p-n junctions protected by a film of silicon dioxide.

transistor with an unprotected surface. The ions and electrons formed beneath the can of the device, during the ionization of the gas, are collected by the electric fields of the junctions on the oppositely charged surfaces. This process can be accelerated or slowed down, depending on the strength and direction of the field between the transistor

171

crystal and the metallic can. In *n-p-n* transistors the electrons migrate to the collector and the positive ions migrate to the base where they form an inversion layer adjacent to the collector, which increases the effective surface of the collector *p-n* junction. This layer can cover the entire base and produce a channel of surface conductivity between the collector and the emitter. Those processes are responsible for the increase in the reverse collector current I_{co} and in the floating potential of the emitter during irradiation.

The recovery of I_{co} after the end of irradiation corresponds to a loss of the charge accumulated on the surface. If there is no bias the ions are not held by their junctions and diffuse along the surface annihilating or shortening the channels and inversion layers. The fact that reapplication of the bias again raises the I_{co}, indicates that the junctions can recapture electrons which, apparently, do not leave the surface. The accelerated recovery in the case of γ-irradiation without a bias indicates that the surface charges are neutralized by the ions and electrons formed in the irradiation.

Those processes also take place if the surface of silicon transistors is protected by an oxide film (Figure 68). The only difference is that the silicon dioxide film, which is a good insulator, prevents a rapid dissociation of the charges accumulated on the surface, in the absence of a bias.

The surface radiation effects can be eliminated, or at least greatly weakened either by evacuating the can or by subjecting it to heat treatment that would prevent the accumulation of charges. Weiss and Knox /248/ found that treating the surface of silicon transistors with a low-melting very pure arsenic-sulfide-iodine glass* produces the same results as evacuation, i.e., greatly reduces the growth rate of I_{co} during γ-irradiation.

The base current transport coefficient, which is a measure of surface processes in transistors, can be stabilized by treating the base surface in a way that would increase the rate of surface recombinations. Such a treatment causes a decrease in the initial value of the base current transport coefficient but radiation changes do not occur until the lifetime of carriers injected into the base change as a result of structural damage in the bulk of the transistors. Experimental studies of transistor irradiation, particularly the results of Brucker /249/ show that such treatment can consist of irradiating the surface or the assembled

---

* 24% As, 67% S, 9% I (by wt.).

transistor with X-rays, electrons, protons or heavier particles which do not change the bulk structure of the crystal.

7. Influence of Operating Conditions and Irradiation Temperature on the Radiation Stability of Transistors

The influence of the operating conditions of transistors on the rate of radiation induced change has been discussed in connection with the influence of radiation effects on the lifetime of minority carriers and on the surface properties of transistors. In this chapter we shall discuss certain practical problems of the operation of transistors under irradiation.

The rate of change of transistor parameters per unit of radiation intensity depends on whether the transistor is active or passive. The relationship between this rate and both the voltage across the junction and the emitter current, depends on whether bulk or surface changes are predominant at a given moment of irradiation. If bulk processes prevail, the dependence of their increase on the collector voltage is determined mainly by the dependence of the effective width of the base region on the potential across the collector p-n junction. The phenomena responsible for that relationship have been discussed above; they are associated with changes in the collector current, which cause a change in the width of the p-n junction and a widening or narrowing of the base region, which determines the rate of increase in the parameters of irradiated transistors.

This effect is most pronounced in high-frequency alloyed transistors and almost absent in high-frequency diffused transistors, particularly in epitaxial transistors in which the broadening of the space charge layer of the p-n junction is mainly in the direction of the collector and not of the base.

It has been found /250/ that if radiation affects mainly the surface processes in transistors, their gain is almost not influenced by the collector bias. However, the bias had a marked effect on the change in the reverse collector current I_{co}. Figure 69 shows that a sudden increase in the collector bias during irradiation /247/ is accompanied by a similar increase in the collector current and by an increase in the slope of the curve representing the dependence of I_{co} on the radiation dose. Those increases are apparently due to the fact that large fields at the p-n junctions facilitate a redistribution of charges on the surface which widens the surface channels and increases the collection of ions from the surrounding space /247/.

173

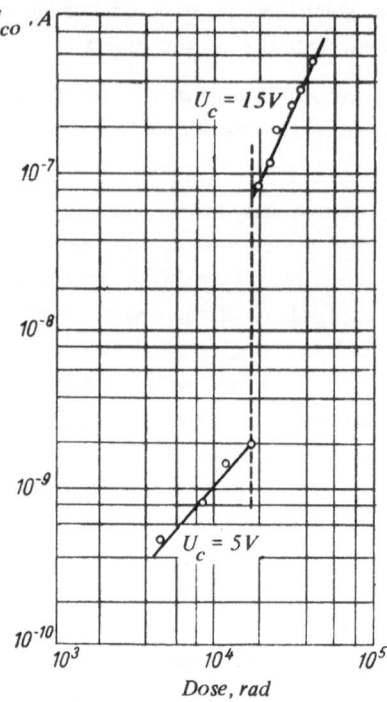

Figure 69. Influence of collector bias on the relationship between the reverse collector current and the γ-radiation dose.

The changes (in the gain of transistors) induced by neutrons and γ-radiation are greatly influenced by the emitter current at which the irradiation and the measurements take place. Figure 70 shows the dependence of the reciprocal of the base current transport coefficient of γ-irradiated 2N1613 *n-p-n* planar diffused silicon transistors on the emitter current /250/. It is evident that an increase in current causes a considerable decrease in the change of the emitter current transport coefficient $(1 - \alpha_\Phi)$. It has been found /72/ that for this type of transistor a current increase from 10^{-6} to 10^{-5} A reduces the rate of change of $(1 - \alpha_\Phi)$ by nearly 50% while a current increase to 10^{-4} A reduces that rate by a further 10% with respect to $d/d\Phi (1 - \alpha_\Phi)$ at 10^{-6} A.

In contrast to the case of γ-irradiated transistors, no such relation exists between the rate of damage production and the emitter current in neutron-irradiated transistors, because of the different nature of the

174

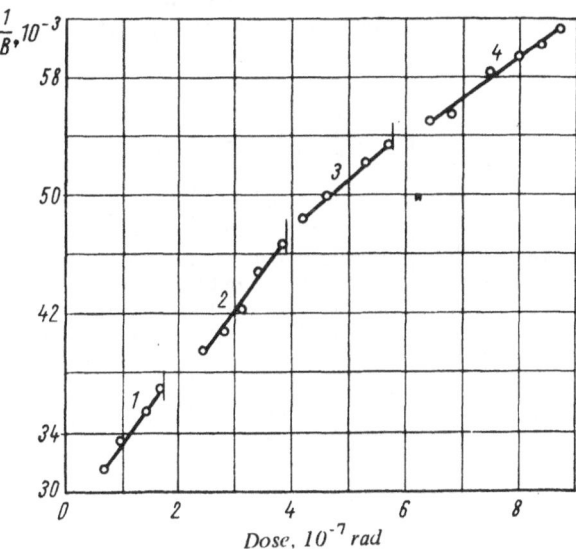

Figure 70. Influence of the operating conditions of transistors on the rate of bulk changes $1/B$, induced by γ-radiation from a Co^{60} source:
1 – no bias; 2 – $U_c = 20\,V, I_c \doteq 10^{-6}\,A$; 3 – 20V, $10^{-5}A$; 4 – 20V, $10^{-4}A$.

radiation defects /251/. Table 10 presents data on neutron irradiation of three types of passive and active *n-p-n* silicon transistors, which confirm the above observation.

The coefficients representing the dependence of lifetime on radiation (K') given in Table 10 were calculated using equation (96) and assuming that the reciprocal of the base current transport coefficient is determined only by the decrease in the lifetime of carriers in the base. In reality the recombination current of the emitter junction and the surface recombination current comprise a considerable if not predominant part of the total base current in transistors operated at low currents /249/. Therefore, the values of K' given in the Table are not real values for the semiconductor material of the base but mean values which take into account the processes in both the *p-n* junction and on the surface.

In order to explain the dependence of damage (in γ-irradiated 2N1613 transistors) on the emitter current, a model has been proposed /250/ according to which the rate of production of stable defects is influenced by the charge of their components. The relative probability

175

Table 10. Comparison of data obtained in neutron-irradiation of transistors in a passive and an active state.

Transistor type	State of transistor during irradiation	Coefficient K', 10^{-6}/RDU·sec at various emitter currents						
		10^{-6} A	10^{-5} A	10^{-4} A	10^{-3} A	10^{-2} A	$3 \cdot 10^{-2}$ A	10^{-1} A
2N1613	Passive	15.3 (9.0)	7.7 (7.7)	3.76 (7.2)	1.92 (6.4)	1.16 (3.3)	0.92 (3.1)	--
2N1613	Active	17.5 (7.7)	6.9 (6.9)	3.31 (6.4)	1.88 (6.6)	1.18 (5.1)	0.96 (4.5)	−
2N916 (Planar epita-xial)	Passive	11.4 (33.0)	4.4 (13.9)	2.21 (11.1)	1.15 (8.4)	0.71 (10.3)	0.52 (14.7)	−
2N916	Active	11.6 (24.2)	4.5 (18.4)	2.11 (15.1)	1.20 (15.0)	0.69 (13.0)	0.51 (16.6)	−
2N708	Passive	18.5 (13.4)	6.1 (5.6)	2.92 (5.6)	1.57 (4.1)	1.06 (2.7)	0.82 (11.5)	0.55 (5.5)
2N708	Active	13.8 (16.2)	5.0 (7.0)	2.46 (5.9)	1.35 (5.8)	0.94 (4.8)	0.77 (5.0)	0.48 (8.3)

Note: During irradiation the operating conditions of the active transistors were as follows: $U_c = 5$V, $I_e = 10^{-3}$A. The numbers in parentheses indicate the relative dispersion in percent. RDU — arbitrary unit of integrated neutron flux which takes into account the spectral effectiveness of the produced damage ("Radiation Damage Unit") /252/.

that annihilation of vacancy-interstitial pairs would take place simultaneously with the formation of complexes that are stable at a given temperature (for instance, vacancy-impurity atoms) is determined by the charges of the interacting components. The charge of those components depends on the conditions of irradiation (temperature, illumination, concentration of non-equilibrium carriers, etc.). Stein /253/, who studied the carrier lifetime in γ-irradiated 6-ohm phosphorus-doped silicon at $76°$K, found that the rate of injection of recombination centers can be increased sixfold by additional illumination of the specimens with light from a tungsten-filament lamp. Similar effects confirming the correctness of that model have been observed by Watkins /59/ in n-type silicon and by Kortright, Klontz, Ishino and Mitchell /254—256/ in germanium.

The above results indicate that the injection of electrons into the base of *n-p-n* silicon transistors, in numbers sufficient to change the charge of components of simple defects, causes a decrease in the rate of formation of stable complexes, which is the predominant process under ordinary conditions. However, such an effect of the emitter current on the rate of formation of radiation changes is not general (for all transistors) but may be different for transistors of different types since it depends on the type of semiconductor, on the type of conductivity, on the concentration and type of the main dopants and of uncontrolled impurities and on the physical conditions under which experiments are carried out. Under identical operating conditions, irradiation of transistors can lead to the intensification of a certain effect, to its weakening or even to its reversal, i.e., to an increase in the rate of damage production during irradiation of transistors in an active state. Irradiation may also not produce any such effects.

The nearly total absence of such an effect in semiconductors and in semiconductor devices irradiated with heavy particles can be attributed to a certain degree to the formation of more complex structural defects, e.g. of clusters.

The great volume of available data on transistor irradiation (see Table 4) indicates that relatively great changes occur in the base current transport coefficient at low currents, regardless of the state of the irradiated transistor. This is illustrated by Figure 71 /251/ which shows the relative dependence of the increase in the reciprocal of the base current transport coefficient on the collector current. The Figure also shows that in the case of irradiation with neutrons the base current transport coefficient is markedly affected by the collector current, (at which the parameters were measured). Genre, Gloten *et al.* /257/ have shown that a similar dependence is found for various *p-n-p* and *n-p-n* silicon transistors irradiated with 14 MeV neutrons, with 2 MeV protons, with 3 MeV electrons and with γ-radiation from a cobalt source. Figures 72 and 73 show the results of Brucker /249/ who irradiated planar *n-p-n* (2N2102) silicon transistors and *p-n-p* (2N1132) transistors with electrons of different energies, with reactor neutrons and with 16.8 MeV protons. Those figures also show that within a certain range of operating currents, which, according to /257/ do not exceed values at which an injection level $x = \Delta n / n_0 = 10^{-1} - 10^{-2}$ is attained, the dependence of the increase $\Delta(1/B)$ on the collector current or on the emitter current can be represented by a straight line

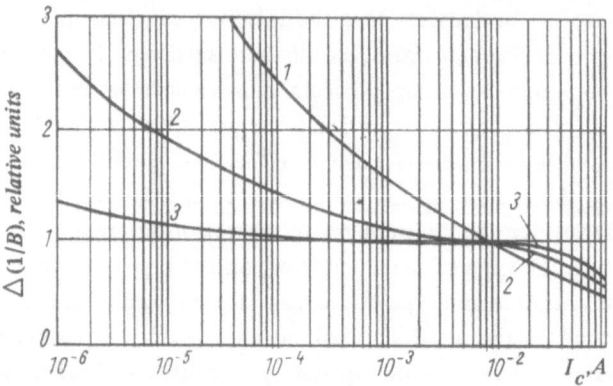

Figure 71. Relative (reduced to a collector current of 10^{-2} A) dependence of the increase in $1/B$ on the collector current at which B was measured for planar 2N1613 transistors irradiated with neutrons in a passive or active state (1), and with γ-rays in an active state (2) and in a passive state (3).

(on a loglog scale) with a slope of -0.25 (-0.3). Thus, equation (97) for current transport coefficients can be revised and written as follows:

$$\Delta(1-\alpha) \approx \Delta(1/B) = Kt_{tr}\,(I_{e0}/I_e)^{0,25}\,\Phi^n. \qquad (144)$$

The current I_{e0} corresponds to an injection level $\Delta n/n_0 = 10^{-2}$ and can be found from:

$$I_{e0} = aA_{pn}^2/I_0\,t_{tr}$$

where $a = 1.73\cdot10^{-18}$ for n-p-n Si and $a = 0.75\cdot10^{-18}$ for p-n-p Si; I_0 can be found experimentally from $I_c = I_0\exp(qU_{eb}/kT)$, A:$A_{pn}$ is the surface area of the emitter junction, cm^2; t_{tr} is the transit time of carriers, sec.

The above dependence of the current transport coefficient of irradiated transistors on the operating current is associated with the fact that if the current is decreased, recombination processes in the space charge layer of the emitter (rather than in the base region) become predominant. This has been confirmed by Goben /258/ and Bruckner /249/ for silicon transistors and by Bilger /199/ for germanium transistors.

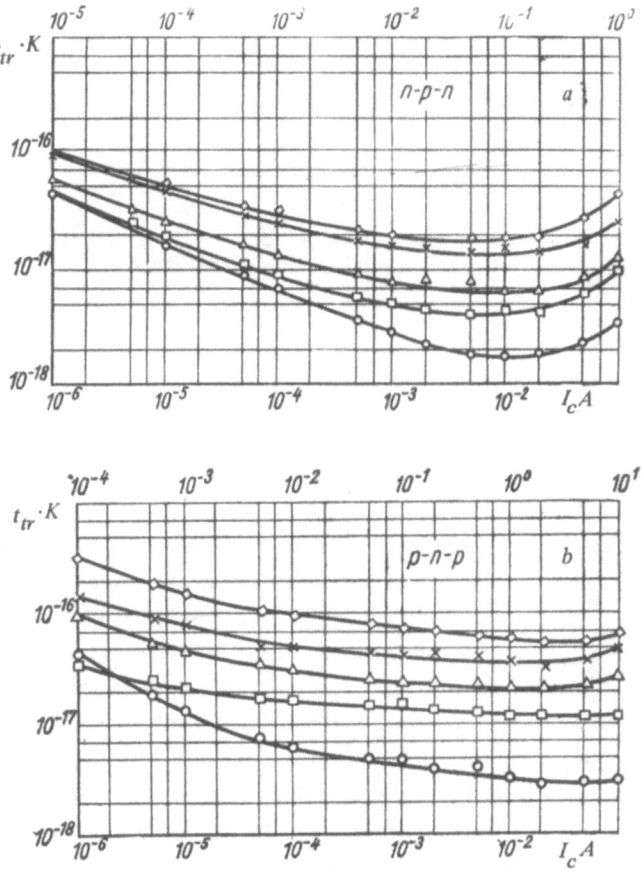

Figure 72. Dependence of the product $t_{tr} \cdot K$ on the collector current for *n-p-n* planar 2N2102 silicon transistors (*a*) and for *p-n-p*-mesa–2N1132 silicon transistors (*b*) irradiated with electrons of various energies: O-275; □-375; △-525; x-725; ◇-1000 KeV.

A similar increase in the change of the current transport coefficient of irradiated transistors has been found at higher injection levels. However, in that case the displacement of the injected current into the peripheral region of the emitter and the broadening of the base of high-frequency diffused transistors also exert a certain influence.

The displacement of the injected current to the periphery of the emitter is caused by a voltage in the base region, directed along the

179

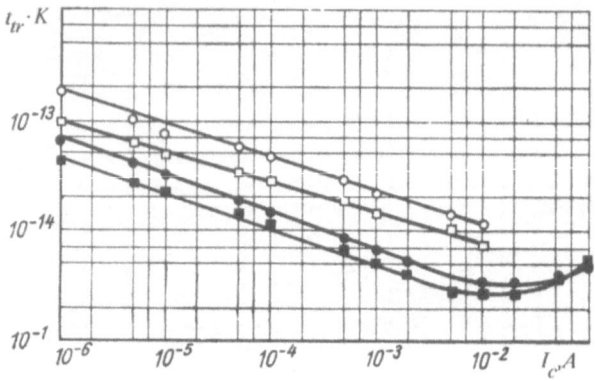

Figure 73. Dependence of the product $t_{tr} \cdot K$ on the collector current in *n-p-n* planar 2N2102 silicon transistors and in *p-n-p* mesa 2N1132 silicon transistors irradiated with 16.8 MeV protons (□ ○) and with reactor neutrons (■ ●).

surface of the *p-n* junction in such a way that the internal region of this junction has a lower potential. The displacement would be more pronounced and would occur earlier if the resistivity of the base material is increased, or if the thickness of the base layer is reduced, or if the recombination of injected carriers in that layer is accelerated. Therefore, in irradiated transistors the displacement would occur at current densities lower than in non-irradiated ones. The displacement of the injected current leads to a weakening of the effect of recombinations in the bulk of the base and to an increase in the influence of surface recombination, as a result of which the effectiveness of the emitter decreases.

Broadening of the base as a result of increasing current density at the collector junction takes place in transistors with a relatively high ohmic collector region and a wide *p-n* junction. This is characteristic in particular for epitaxial silicon transistors. If the current density of charge carriers from the base is high, their concentration in the epitaxial collector layer can increase to such a degree that that layer will function as an extension of the base region rather than as a collector. This effect becomes stronger as the voltage across the collector junction is reduced /190/.

Base broadening may occur in non-irradiated transistors, too. However, in that case it is manifested mainly as a deterioration of

180

high-frequency properties rather than as a total decrease in the base current transport coefficient. If irradiation causes an increase in the resistance of the collector and the epitaxial layer, then base broadening would take place at lower current densities.

The temperature dependence of the increase in the change in the gain of transistors is also of interest. Thus, γ-irradiation of 2N1613 transistors at about 40°C causes $\sim$ 10% greater residual damage than irradiation at 60°C /250/. The transistor parameters were in both cases measured at 60°C. If the parameters are measured at the temperature of the experiment, the difference reaches 20%.

Such temperature dependence is due to annealing of radiation-induced structural defects; therefore, an increase in temperature leads to a decrease in the concentration of the stable defects and in the change caused by unit radiation dose.

The nature and rate of defect annealing is determined by the type of defects, i.e., by the nature of irradiation. Therefore, the temperature effects of neutron and γ-irradiation are different. Those effects depend also on the carrier injection level. The lower that level, the greater is the influence of the irradiation temperature on the final results. The temperature effect is greater in "exhausted" transistors. Figure 74,

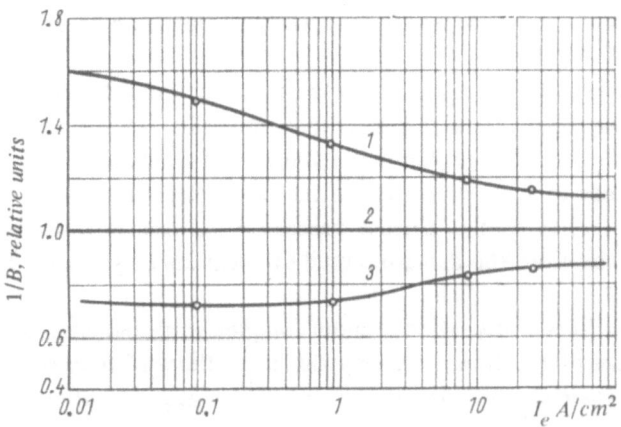

Figure 74. Relative (reduced to T=35°C) dependence of the change in $1/B$ on the emitter current density for passive 2N1613 transistors irradiated at various temperatures: $1-10°$; $2-35°$; $3-60°$C.

which shows the effects of neutron irradiation on 2N1613 transistors at different temperatures, illustrates the relative dependence of that effect

181

(reduced to 35°C) on the emitter current density /251/.

It is now clear that all phenomena in irradiated transistors, regardless of transistor type, structure and material, are associated with processes both on the surface and in the bulk of the semiconductor crystal.

Surface processes may cause substantial changes in the basic parameters (base current transport coefficient B and particularly the reverse collector current I_{co}). However, since such processes take place at relatively small radiation doses and rapidly reach saturation, they cannot be a predominant factor. The nature of such processes and the changes they induce have not yet been correlated with the kind, intensity and dose of radiation on one hand and with the initial parameters of the transistors on the other. This could be attributed to the fact that the surface properties are greatly influenced by the state of the crystal surface, by the composition of the surrounding atmosphere, by the tightness of the protective can and by many other factors.

Nevertheless, it may be expected that effective methods would be developed for the protection of the surface and of the transistors, which will reduce to a minimum the influence of ionizing radiation on the state and properties of semiconductor surfaces. Some methods, such as evacuation of the protective can and advance irradiation with "soft" radiation have already been developed and can be successfully used even at present.

The problem of processes taking place in the bulk of the crystals, which damage the crystal lattice, is more difficult. Since it is almost irreversible, such damage limits the lifetime of irradiated transistors. The influence of bulk processes on the parameters and operation of transistors (unlike that of surface processes) has been investigated in detail and therefore the damage caused by them can be predicted fairly accurately.

The principles of operation of more complex semiconductor devices, among them of thyristors, are very similar to those of transistors and we shall therefore discuss below in brief their operation under irradiation.

8. Some Problems of Thyristor Operation Under Irradiation

Thyristors are four-layer semiconductor devices with a *p-n-p-n* structure. A thyristor circuit is shown in Figure 75. Without dealing in detail with the operating principles of thyristors (which have been

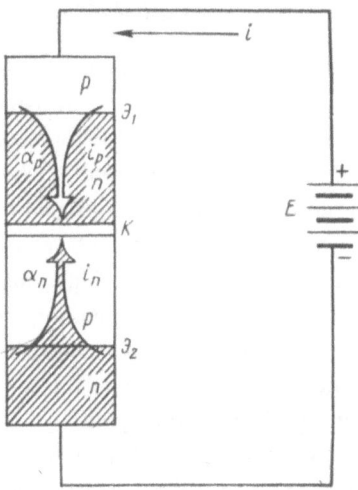

Figure 75. Structure and circuit of a thyristor.

discussed in the Soviet literature /259, 260/),we shall only point out the basic principles of their operation, needed for understanding the problems.

If a bias is applied on a thyristor as shown in Figure 75, the terminal p-n junctions, which shall henceforth be called emitters, will be forward biased while the middle junctions (collectors) would be reverse biased. The emitters can inject minority carriers into the base layers, and those carriers diffuse into the middle collector junction. Thus, the transport of carriers in thyristors is similar to that in ordinary transistors. Therefore the transfer processes may be characterized by the hole current transport coefficient in an n-type base layer α_p and of an electron current in a p-type base layer α_n. The motion of electrons and holes through the collector junction can lead to avalanche multiplication which is characterized by the factors M_n and M_p, If we designate by i the total current through the device, by i_0 the reverse current of the collector junction and by U_c the voltage on the collector junction (which is equal to the external voltage applied to the thyristor with an accuracy up to the sum of the voltages on the emitters) we can write:

$$i = \frac{i_0 (e^{-\frac{qU_c}{mkT}} - 1)}{1 - \alpha_p M_p - \alpha_n M_n} =: \frac{i_0 (e^{-\frac{qU_c}{mkT}} - 1)}{1 - \alpha_t} , \qquad (145)$$

183

where

$$\alpha_t = \alpha_p M_p + \alpha_n M_n.$$

Equation (145) indicates that the condition $\alpha_t = 1$ corresponds to a break on the V–I curve of thyristors, beyond which there is a region with negative resistance which is responsible for its switching capability.

The condition $\alpha_t = 1$ is fulfilled, apparently, when $\alpha_p M_p \approx 0.5$ and $\alpha_n M_n \approx 0.5$. In the absence of any control signal the switching voltage U_{sw} of real devices is always smaller than the value at which avalanche multiplication begins. Therefore, switching may occur only if $\alpha_p + \alpha_n = 1$. If the current through the junctions is small, the current transport coefficient is almost entirely determined by the emitter effectiveness, γ, which, in turn, depends on the voltage across the junction or on the current density through it.

The switching voltage is controlled externally by passing an additional current through one of the emitter junctions as a result of which both α_p and α_n are raised to the level necessary for switching.

Thus, the operation of thyristors is based on the current dependence of the current transport coefficients α_n and α_p or, more accurately, on the current dependence of the capability of emitters to inject minority carriers into base layers.

Knowing the influence of radiation on the emitter effectiveness at low current densities, we can establish (at least qualitatively) the influence of radiation on the basic parameters and the V–I characteristics of thyristors.

As we showed above, the considerable influence of radiation on the lifetime of carriers is the main factor that causes a decrease in the effectiveness of emitters at low current densities. Therefore, in order to switch a thyristor, the current density at its emitters must continuously increase during irradiation. In real cases one of the current transport coefficients is always much greater than the other. Therefore, during irradiation, changes in that coefficient would determine the changes in thyristor parameters, among them the switching current at constant collector voltage.

In the absence of a control current the increase in emitter current density, necessary for switching, will correspond to an increase in the switching voltage U_{sw}, since an increase in the collector voltage causes increases in the reverse current. At very high radiation levels, when the effectiveness of the base emitter decreases so much that even at maximum collector voltage the reverse current will be insufficient for

184

switching, switching could occur at a voltage U_c at which avalanche multiplication of carriers takes place, i.e. at breakdown voltage.

The decrease in emitter effectiveness as a result of irradiation may also cause an increase in the switching current of the control electrode at a constant voltage on the device. In addition, a decrease in $d\alpha/di_{p(n)}$ at currents at which $\alpha_t = 1$ will produce a "soft" switching characteristic for the blocked state region, which corresponds to a gradual approach to the negative-resistance section on the V–I curve.

Those changes in the parameters and in the V–I curve of thyristors, which have been established on the basis of their operating mechanism and the effect of radiation on transistors, have been confirmed experimentally. Figure 76 shows the characteristics of low-power

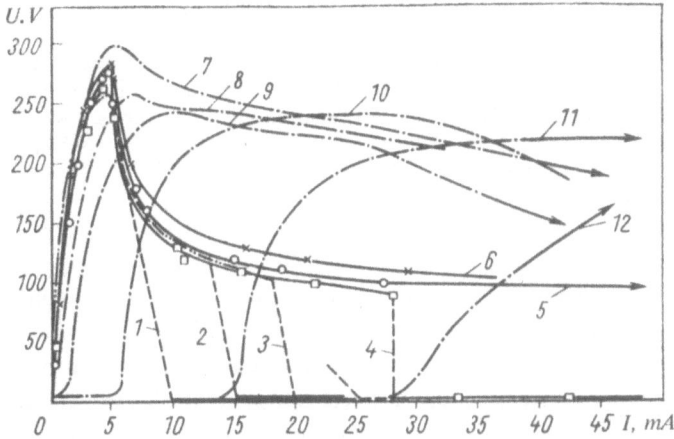

Figure 76. V–I characteristics of a low-power thyristor before and after irradiation with various integrated fast-neutron doses.

$1 - I_c = 0 \ \Phi = 0; 2 - I_c = 0 \ \Phi = 3{\cdot}10^{11} \ cm^{-2}; 3 - I_c = 0 \ \Phi = 10^{12} \ cm^{-2};$
$4 - I_c = 0 \ \Phi = 3{\cdot}10^{12} \ cm^{-2}; 5 - I_c = 0 \ \Phi = 10^{13} \ cm^{-2}; 6 - I_c = 0 \ \Phi =$
$= 3{\cdot}10^{13} \ cm^{-2}; 7 - I_c = 0 \ \Phi = 10^{14} \ cm^{-2}; 8 - I_c = 5 \ mA \ \Phi = 10^{14} cm^{-2};$
$9 - I_c = 10 \ mA \ \Phi = 10^{14} \ cm^{-2}; 10 - I_c = 20 \ mA \ \Phi = 10^{14} \ cm^{-2}; 11 - I_c =$
$= 50 \ mA \ \Phi = 10^{14} \ cm^{-2}; 12 - I_c = 90 \ mA \ \Phi = 10^{14} \ cm^{-2}.$

thyristors at various control currents and integrated fast neutron doses; it indicates that an increase in radiation dose leads to an increase in the switching voltage at zero control current and that at the "blocked" region the characteristics are very "soft", particularly at high control currents.

Figure 77 shows the dependences of one of the current transition coefficients on the current flowing through a junction of the 2N680

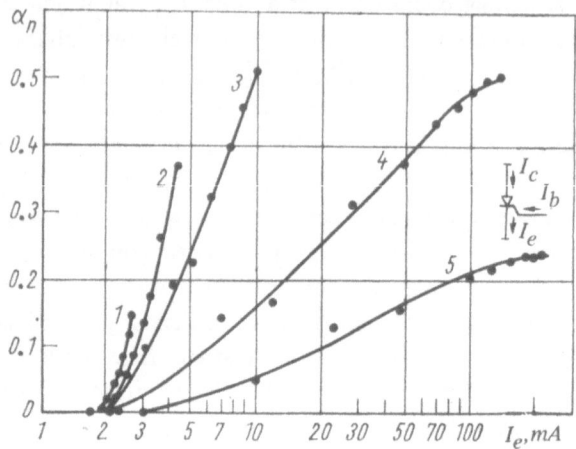

Figure 77. Dependence of the current transport coefficients a_n of a controlled 2N680 type diode on the current before and after irradiation with various fast neutron integrated fluxes. $1 - \Phi = 0$; $2 - \Phi = 9 \cdot 10^{11} \text{cm}^{-2}$; $3 - 6.3 \cdot 10^{12} \text{cm}^{-2}$; $4 - 2.25 \cdot 10^{13} \text{cm}^{-2}$; $5 - 4.25 \cdot 10^{13} \text{cm}^{-2}$.

low-power control diode /261/ (General Electric), at various integrated neutron doses; the control current necessary for switching of semiconductor devices under irradiation is very high.

Let us consider another important parameter of thyristors — the saturation voltage, i.e. the voltage across an open thyristor with all junctions forward biased. Such thyristors are similar to transistors in a state of saturation or to diodes in which the conductivity of the base material is modulated only by the injected carriers. An analysis of diode operation shows that the voltage drop on diodes operating under such conditions increases rapidly under irradiation. The same is true for the saturation voltage of thyristors. Leith /261/ found that in low-power devices of the 2N680 type the voltage drop can reach tens or hundreds of volts at integrated neutron fluxes of about 10^{13} cm^{-2}. This means that such neutron doses cause complete loss of the switching properties of low-power thyristors.

CHAPTER II

RADIATION EFFECTS IN SEMICONDUCTOR DIODES

The properties of semiconductor diodes, like those of other semiconductor devices are greatly influenced by irradiation. Thus, both branches of the V–I curve are changed. However, the magnitude of those changes depends on the type of semiconductor, on the design of the device and on the operating conditions. Accordingly, the radiation stability of diodes is sometimes determined by the degree of deformation of the forward V–I characteristics and sometimes by the changes in reverse characteristics. Therefore, in contrast to transistors, it is difficult to introduce a general criterion for the radiation stability of diodes.

Usually, in the case of germanium diodes for which the main changes are in the reverse branch, the radiation stability criterion is either the maximum permissible saturation current or its permissible relative increase. For silicon diodes the criteria of radiation stability are also certain maximum permissible changes, either in the voltage drop on the forward branch, or the forward resistance, since its changes by radiation are characterized by the change in the forward V–I characteristics.

At high radiation doses semiconductor diodes lose their main feature, i.e., that of being a rectifying device, and become linear semiconductor resistors with a low (germanium) or high (silicon) resistivity, regardless of whether the radiation changes are predominant in the forward or reverse branches.

In spite of the apparent simplicity of diode structure, compared with that of transistors or other devices with several *p-n* junctions, the processes taking place in diodes under irradiation (which cause the above changes) are not less complex. The rate and degree of radiation changes in all semiconductor devices are not limited to *p-n* junctions but involve the properties of the base region, of the ohmic contact and the surface effects.

Today much attention is given to processes taking place in irradiated diodes and to the factors that determine the radiation stability of semiconductor devices.

1. Stable Radiation-Induced Changes in the Forward Branch of the V−I Characteristics of Diodes.

As we mentioned above, irradiation of silicon diodes affects mainly the forward V−I characteristics. An increase in the integrated radiation flux causes a monotonous decrease in the forward voltage and differential resistance /187, 262/. Ultimately, the silicon diode tends to become a linear high-ohmic resistor with almost no non-linearity of diodes during polarity changes. Figure 78 shows an oscillogram of the

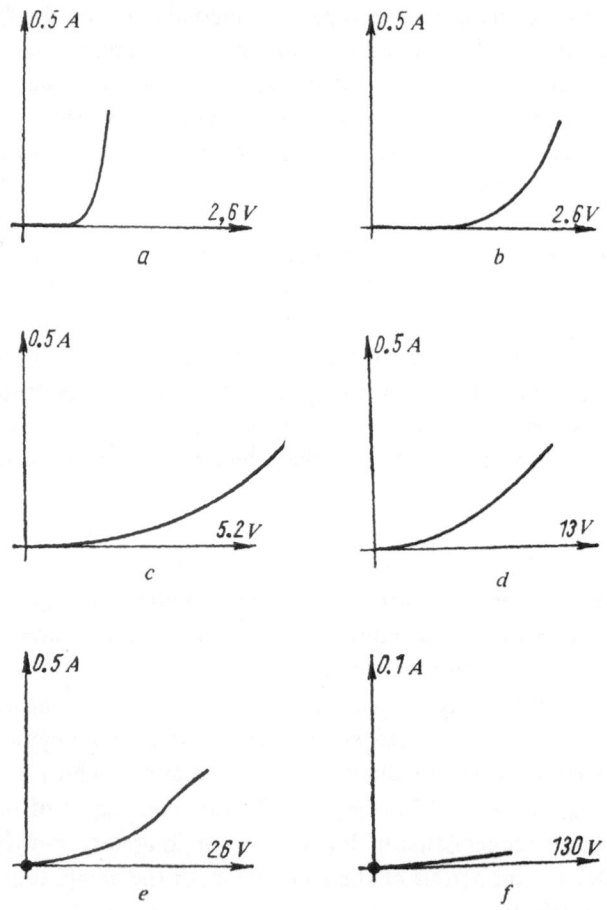

Figure 78. Oscillogram of the forward V−I characteristics of a medium-power silicon rectifying diode after irradiation with various integrated fast neutron fluxes: a-0; b-10^{13}; c-$3 \cdot 10^{13}$; d-$5 \cdot 10^{13}$; e-10^{14}; f-$3 \cdot 10^{14}$ cm^{-2}.

188

forward characteristics of a silicon rectifying medium-power diode, which illustrates the nature of radiation-induced changes.

In order to evaluate the role of the p-n junction and of the base region in the changes of forward characteristics, let us represent the total voltage drop on the diode as a sum of the voltages across the p-n junction U_{pn} and across the base U_b.

$$U_d = U_{pn} + U_b. \qquad (146)$$

The dependence of the V–I characteristics of the intrinsic p-n junction on the applied potential is governed mainly by processes which take place either in the base region of the diode or in the p-n junction. Those characteristics also depend on the properties of the back ohmic contact if the diffusion length of charged minority carriers in the base L_b is greater than the thickness of the base w_b.

Sah *et al.* /198/ found that in silicon diodes operating at forward potentials not exceeding $8kT/q$, the component contributed by recombination of carriers in the space charge layer is predominant in the current flowing through the p-n junction. The dependence of the voltage U_{pn} on the total current density in this range (J_{pr}) can be written:

$$U_{pn} = \frac{mkT}{q} \ln \left[\frac{J_{pr} \tau_{pn}}{q n_i w_{pn}} \right] , \qquad (147)$$

where m has a value between 1 and 2, depending on the value of U_{pn} and on the position of recombination levels in the forbidden gap:

$$\tau_{pn} = \sqrt{\tau_{p0} \cdot \tau_{n0}},$$

w_{pn} is the width of the space charge region.

At forward p-n junction voltages greater than $10kT/q$, the diffusion component (which is determined by the propagation and recombination of carriers injected into the base from the p-n junction) is the predominant part of the total current. This refers to diodes in which the p-n junction (gradual or abrupt) consists of regions with greatly differing concentrations of majority carriers. In such cases, the voltage U_{pn} depends both on the w_b/L_b ratio and on the properties of the

189

ohmic contact. If the height of the potential barrier on the ohmic contact, regardless of its type (depleted or not-depleted) does not exceed several kT, then the voltage drop will not exceed kT/q and the nonequilibrium concentration of minority carriers would remain lower than their equilibrium concentration at that contact. Thus, for diodes with n-type conductivity and a strongly doped p-region, $\Delta p_n(w_b) < p_n$. By solving the continuity equation for the base region of such semiconductors in a univariate approximation (and taking into account the above limiting condition for the ohmic contact) we obtain:

$$U_{pn} = \frac{kT}{q} \ln\left[\frac{J_{pr} w_b}{q p_n D_p} \cdot \frac{\text{th}\,(w_b/L_p)}{w_p/L_p}\right].$$ (148)

The upper validity limit of equation (148) is determined by the condition of a low injection level $\Delta p_n(0) \leqslant n_n$.

For high injection levels, i.e. $\Delta p_n(0) \gg n_n$, the term U_{pn} can be written either in the form suggested by Sah *et al.* /198/:

$$U_{pn} = \frac{2kT}{q} \ln \frac{J_{pr} L_0}{2 q n_i D_0},$$ (149)

where

$$L_0 = \sqrt{D_0 \tau_\infty} = \sqrt{\frac{2b}{b+1} D_p \tau_{p0} + \tau_{n0}}$$

is the bipolar diffusion length, or (as suggested by V.I. Stafeev /263/) for "elongated" diodes where $w_b/L_b \gg 1$ we could write:

$$U_{pn} = \frac{ckT}{q} \ln\left[\frac{2 J_{pr}\,(b+1)\,L_0}{q b n_n D_0} \cdot \frac{1}{\text{ch}\,(w_b/L_0)}\right]$$ (150)

and for diodes for which $w_b/L_b \leqslant 1$ we have:

$$U_{pn} = \frac{2kT}{q} \ln \frac{J_{pr} w_b}{2 q D_p n_i}.$$ (151)

Those equations for U_{pn} can be reduced to:

$$U_{pn} = \frac{mkT}{q}\left[\ln\frac{J_{pr}}{q\Pi} + \chi \ln \tau\right],$$ (152)

where τ is the lifetime of minority carriers either in the space charge field (τ_{pn}) or in the base region of the diode at small (τ_p) and large ($\tau_{p\infty} = \tau_{po} + \tau_{no}$) injection levels and Π is the product of all other logarithmic terms in equations (148–151) which can be considered constant during their irradiation, compared with the lifetime τ. In equation (152) $1 \leqslant m \leqslant 2$* and $0 \leqslant \chi \leqslant 1$.

The lifetime τ monotonically decreases during irradiation and therefore the voltage U_{pn} must also decrease. The rate of that decrease V/(neutron/cm^2) can be found by differentiation of (152) with respect to Φ:

$$\frac{dU_{pn}}{d\Phi}\bigg|_{J_{pr}=\text{const}} = -\frac{mkT}{q}\chi\tau K, \qquad (153)$$

where K is a coefficient for the radiation effect on the lifetime (τ_{pn}, τ_p or τ_∞). Equation (153) shows that the radiation changes in U_{pn} would be the smallest in diodes with a thin base ($w_b/L_b \leqslant 1$) at medium and high injection levels, when the total current through the p-n junction is determined by diffusion rather than by the recombination component.

The voltage drop at the base region of the diode (w_b) depending on the operating conditions is determined either by the conductivity of the base material, or by the degree of its modulation by minority carriers injected from the p-n junction. At low injection levels, i.e., if $\Delta p(0) < n_n$, we have:

$$U_b = J_{pr} \cdot w_b / \sigma_n \qquad (154)$$

but if $w_b \leqslant (2-3)L_b$, U_b is not larger than a few times kT/q. However, if the p and n regions of the diodes are thick or the integrated radiation fluxes (doses) are large then even diodes with an initially thin base do not satisfy the condition $w_b/L_b \leqslant 1$, because of a decrease in the lifetime and diffusion length of minority carriers so that U_b would increase and approach U_{pn}. In such cases the radiation change in U_b is quite rapid, i.e.,

---

* Sah has found that in real diodes m can be larger than 2 as a result of surface phenomena /200/.

$$\left.\frac{dU_b}{d\Phi}\right|_{J_{pr}=\text{const}} = -\frac{J_{pr} \cdot w_b}{\sigma_n^2} \cdot \frac{d\sigma_n}{d\Phi} = -U_{b0}\frac{1}{\sigma_n} \cdot \frac{d\sigma_n}{d\Phi} \qquad (155)$$

The conductivity of the most widely used semiconductor materials (silicon and germanium) decreases upon irradiation at room temperature. High-resistivity p-type germanium is an exception and its conductivity may increase to a certain limiting value. Therefore the voltage U_b, unlike U_{pn}, would usually increase monotonically with irradiation. According to equation (155) the rate of that increase is determined by the initial conductivity of the material τ_n, by the rate of its radiation change $d\sigma_n/d\Phi$, and by the initial value of U_{b0} and consequently by J_{pr}.

Since the radiation changes in U_{pn} and U_b are in opposite directions, diodes with a relatively thick base, i.e. $\chi = 0$, and diodes with a low-resistivity base have a point on the forward V–I characteristics where the rate of the potential drop U_d (caused by irradiation) will be equal to zero at least during the first stage of irradiation. For that point we could write:

$$\frac{mkT}{q}\chi\tau K + U_{b0}\frac{1}{\sigma_n} \cdot \frac{d\sigma_n}{d\Phi} = 0, \qquad (156)$$

and hence

$$J_{pr}^{\circ} = -\frac{mkT\chi\tau K\sigma_n^2}{qw_b \, d\sigma_n/d\Phi} \qquad (157)$$

When $J_{pr} < J_{pr}^0$ the voltage U_d would decrease upon irradiation while in the case of $J_{pr} > J_{pr}^0$ U_d would increase. Such behavior of diodes is observed only within a certain range of radiation doses, outside which the "critical" point on the V–I characteristics would shift towards lower current densities, since neither K nor $d\sigma_n/d\Phi$, remain constant under irradiation. Nevertheless, equation (157) may be valuable for the development of diodes with a high radiation stability. Silicon diodes, in particular low-power ones, are used at current densities much higher (one order of magnitude or more), than the limiting value of J_{pr}^0 but still below such high injection levels that cause modulation of the base material conductivity. Under such conditions the radiation changes in forward voltage (and differential forward resistivity) would be determined solely by changes in the conductivity of the base material.

The dependence of the resistivity of silicon on the integrated radiation flux or dose $\sigma(\Phi)$, can be represented by an exponential function:

$$\sigma(\Phi) = \sigma_0 \exp(-K_\rho \Phi). \tag{158}$$

Using equations (154) and (158) and disregarding voltage variations on the p-n junction, the relative increase in the forward voltage drop on the diode at constant current can be found from:

$$\frac{\Delta U_d(\Phi)}{U_{do}} = \frac{1}{a}(e^{K_\rho \Phi} - 1), \tag{159}$$

provided that in the initial state we have:

$$U_{bo} = (1/a)U_{do}. \tag{160}$$

where
$$a = \frac{J_B}{J_{pr}} \ln\left[b\frac{J_{pr}}{J_B}\left(\frac{n_n}{n_i}\right)^2\right] + 1.$$

Here J_B designates the current density of ohmic degeneracy as defined by Stepanenko /194/:

$$J_B = \frac{kT}{q} \cdot \frac{\sigma_b}{w_b}. \tag{161}$$

The permissible change in the forward voltage U_d of diodes depends on the operating conditions. However, doubling the forward voltage drop at a constant current is often used as a "universal" criterion of their performance. If for a given current $a = 2$, then the maximum integrated radiation flux $\Phi_{2.0}$ which meets the above criterion would be: $e^{K_\rho \Phi_{2.0}} = 5$ or $\Phi_{2.0} = 1.61/K_\rho$.

Since $\left.\dfrac{d\sigma}{d\Phi}\right|_{\Phi \to 0} = -K_\rho \sigma_0 e^{-K_\rho \Phi}$, the coefficient K_ρ in the exponent would be equal to the initial relative rate of change in the conductivity of the material:

$$K_\rho = \frac{d\sigma(\Phi)/d\Phi}{\sigma(\Phi)} \qquad \text{at} \quad \Phi \to 0 \qquad (162)$$

or to the relative rate of elimination of majority carriers, since at small Φ their mobility is constant (μ = constant).

Consequently,

$$K_\rho = \frac{dn/d\Phi}{n} \qquad \text{at} \quad \Phi \to 0 \qquad (163)$$

for n-type material and, correspondingly

$$K_\rho = \frac{dp/d\Phi}{p} \qquad \text{at} \quad \Phi \to 0 \qquad (164)$$

for p-type material.

Taking equations (162) and (164) and the equation for $\Phi_{2.0}$ we obtain:

$$\Phi_{2.0} = 1.61 \frac{\sigma_{n0}}{q\mu_n \left.\frac{dn_n}{d\Phi}\right|_{\Phi \to 0}} \qquad (165)$$

for n-type material and by analogy:

$$\Phi_{2.0} = 1.61 \frac{\sigma_{p0}}{q\mu_p \left.\frac{dp_p}{d\Phi}\right|_{\Phi \to 0}} \qquad (166)$$

for p-type material. Equation (166) has been derived on the basis of an experimental determination of the rate of hole removal and is shown graphically in Figure 79 /264/. Equations (165) and (166) and the data from Figure 79 show that a well-defined relation exists between the starting conductivity of the base material of diodes σ_0 and their radiation stability whose criterion is the integrated radiation flux $\Phi_{2.0}$ corresponding to a doubling of the forward voltage. It was found that all other conditions being equal, the radiation stability of diodes is proportional to the starting conductivity of the material.

The initial conductivity of the material is a parameter that represents the distribution of the dopant in the p-n junction, and the reverse

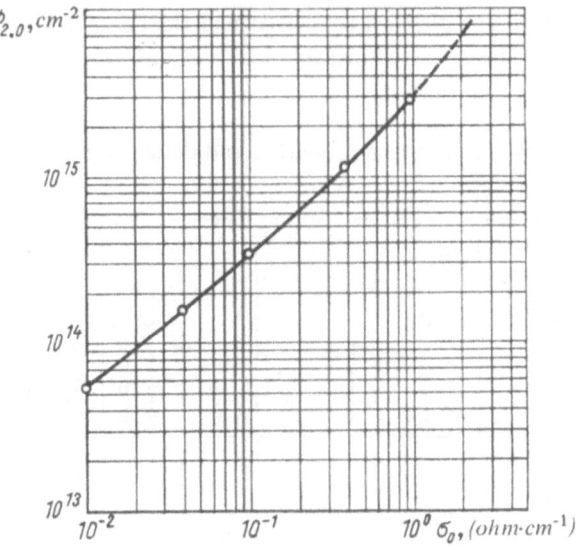

Figure 79. Effect of the conductivity of p-type silicon on the integrated fast neutron flux at which the forward voltage drop is double the inital $\Phi_{2.0}$ value.

breakdown voltage of diodes should be related to their radiation stability. The existance of such a relation has been proved experimentally (Figure 80) /264/. Figure 80 shows the dependence of the relative

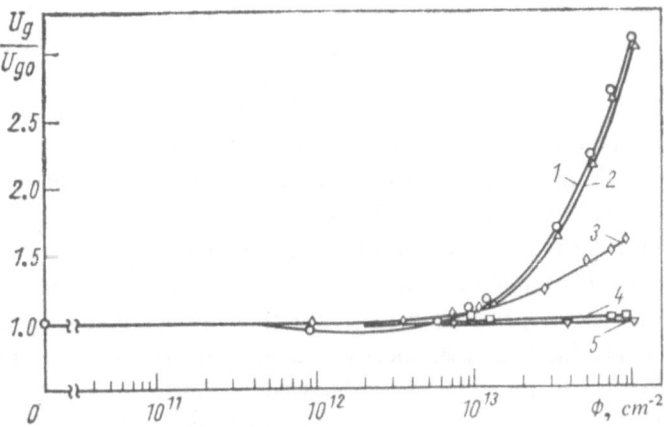

Figure 80. Dependence of the relative change in the forward voltage drop in silicon diodes on the integrated fast neutron flux at energies above 0.01 MeV. The numbers on the figure correspond to the designation of diodes in Table 11.

change in the forward voltage of different diodes on the integrated neutron flux. The physical properties and the design factors associated with such diodes are listed in Table 11.

Table 11. Physical properties and design factors of diodes.

	Diode type	Base material		w_b, μ	w_{pn}, μ	Surface area of $p\text{-}n$ junction, mm^2	$p_p \cdot K_\rho$ neutron·cm^{-1}	$U_{breakdown}$, volt	J, amp/cm^2 (during-irradiation)
		Type of conductivity	resistivity ohm·cm						
1	1N673	p	20	180	33	1.02	5.4	620	4.87
2	1N947	p	60	180	33	1.02	4.1	750	4.87
3	1N2791	p	25	180	33	0.033	5.4	464	3.08
4	1N696 *	n	0.15	180	20	0.016	—	37	3.96
5	1N697 *	p	40	150	43	0.46	4.8	230	4.38

*The material of the base of these diodes was doped with gold.

The base thickness w_b is another factor that determines the stability of silicon diodes. The role of that factor would not be discussed here since it is obvious from the above equation for U_{pn} and U_b. We would only point out that an increase in the radiation stability of diodes due to a decrease in the thickness of their base layer causes also a decrease in their reverse breakdown voltage. Therefore a semiconductor with a high radiation stability cannot have a high breakdown voltage.

At high injection levels when the concentration of charged minority carriers exceeds the concentration of majority carriers in the base region, the voltage U_b is affected by modulation of the conductivity of the material by minority carriers. This is the case with most rectifying devices and low-power diodes and therefore an analysis of the processes taking place in their bases under irradiation is of great practical interest.

The degree and depth of conductivity modulation is determined by the injection levels, by the nature of the back resistivity of the base contact and by the effective diffusion length of minority carriers. This length depends on the injection level since the lifetime in the material is determined by the concentration ratio of minority to majority carriers.

At high minority carrier concentrations, such carriers diffuse and their excess charge is compensated by redistribution of the charge of majority carriers. However, this does not interfere with diffusion processes whose rate is determined by the concentration gradient. Nevertheless, the diffusion coefficients differ, under such conditions, from the diffusion coefficient of holes D_p and of electrons D_n although there is a correlation between them:

$$D_0 = \frac{2b}{b+1}\, D_p = \frac{2}{b+1}\, D_n. \tag{167}$$

The coefficient D_0, unlike D_p and D_n, is usually known as the bipolar diffusion coefficient.

The effective diffusion length L_0 would then be equal to:

$$L_0 = \sqrt{D_0 \tau_\infty} = \sqrt{D_0 \left(\tau_{p0} + \tau_{n0}\right)}. \tag{168}$$

The voltage drop on the base U_b can be found by integrating the lifetime $E(w)$ with respect to the base thickness w_b. An equation for $E(x)$ for n-type semiconductors can be written:

$$E(x) = \frac{J_{pr}}{q\mu_p \left[(b+1)\, p(x) + bn_n + p_n\right]} - \\ - \frac{D_p(b-1)\dfrac{dp(x)}{dx}}{\mu_p \left[(b+1)\, p(x) + bn_n + p_n\right]} \tag{169}$$

(J_{pr} – forward current density).

Usually $n_n \gg p_n$ and therefore whenever possible we should disregard the equilibrium concentration of minority carriers.

Let us consider a purely ohmic, back contact; the recombination rate of minority carriers on it is $s(w) \to \infty$. The redistribution of minority carriers injected by the p-n junctions into the base $p(x)$ can be calculated from /194/:

$$p(x) = \Delta p(0) \, \frac{\text{sh}\,\dfrac{wb - x}{L_0}}{\text{sh}\,(w_b/L_0)} . \tag{170}$$

After integrating equation (169) and taking into account equation (170) we obtain:

$$U_b = \frac{J_{pr} w_b}{\sigma_n} \cdot \frac{2B}{(w_b L) \sqrt{1+B^2}} \, \mathrm{arth} \, \frac{\sqrt{1+B^2} \, \mathrm{th}\left(\frac{w_b}{2L}\right)}{B + \mathrm{th}\left(\frac{w_b}{2L}\right)} +$$

$$+ \frac{kT}{q} \frac{(b-1)}{(b+1)} \ln \left(\frac{\mathrm{sh}\left(\frac{w_b}{L}\right)}{B} + .1 \right) \tag{171}$$

However, B designates a value $b/(b+1)$ times smaller than the reciprocal of the injection level $x = \Delta p(0)/n_n$; it can be readily expressed by means of the forward current density J_{pr}, by the design parameters and by the properties of the semiconductor material:

$$B = \frac{b}{b+1} \cdot \frac{n_n}{\Delta p(0)} = \frac{2b}{(b+1)^2} \cdot \frac{J_B}{J_{pr}} \left(\frac{w_b}{L}\right) \mathrm{ch}\left(\frac{w_b}{L}\right) =$$

$$= \frac{2b^2}{(b+1)^2} \cdot \frac{q D_p n_n \left(\frac{w_b}{L}\right) \mathrm{ch}\left(\frac{w_b}{L}\right)}{J_{pr} w_b} \tag{172}$$

However, as before, $J_B = \frac{kT}{q} \frac{\sigma}{w_b}$

By rearranging equation (171), and taking (172) into account, we obtain:

$$U_b = \frac{4kT}{q} \cdot \frac{b}{(b+1)^2} \frac{\mathrm{ch}\left(\frac{w_b}{L}\right)}{\sqrt{1+B^2}} \, \mathrm{arth} \, \frac{\sqrt{1+B^2} \, \mathrm{th}\left(\frac{w_b}{2L}\right)}{B + \mathrm{th}\left(\frac{w_b}{2L}\right)} +$$

$$+ \frac{kT}{q} \cdot \frac{b-1}{b+1} \ln \left(\frac{\mathrm{sh}\left(\frac{w_b}{L}\right)}{B} + 1 \right) . \tag{171a}$$

In equation (171a) the first term represents the voltage drop on the base layer, modulated by the injected carriers. The second term represents the Dember e.m.f. which is due to a concentration gradient of carriers in the base and to the difference between the mobilities of electrons and holes.

At low injection levels ($B \gg 1$) the second term approaches zero while the first term is transformed into an expression determined by (154):

$$U_b = \frac{4kT}{q} \cdot \frac{b}{(b+1)^2} \cdot \frac{\mathrm{ch}\left(\frac{w_b}{L}\right)}{B} \, \mathrm{arth} \left[\mathrm{th}\left(\frac{w_b}{2L}\right) \right] = \frac{J_{pr} w_b}{\sigma}. \tag{172a}$$

For the other extreme case ($B \ll 1$) we obtain:

$$U_b = \frac{2kT}{q} \cdot \frac{b}{(b+1)^2} \, \mathrm{ch}\left(\frac{w_b}{L}\right) \ln\left(1 + \frac{2\,\mathrm{th}\,\dfrac{w_b}{L}}{B}\right) +$$

$$+ \frac{kT}{q} \cdot \frac{b-1}{b+1} \ln\left(1 + \frac{\mathrm{sh}\,\dfrac{w_b}{L}}{B}\right) \tag{172b}$$

If the conductivity of the base is modulated, the rate of radiation-induced change in the voltage U_b can be found, as before, by differentiating equation (170) with respect to Φ:

$$\frac{dU_b}{d\Phi} = \frac{2\,J_{pr}w_b B}{y\,\sqrt{1+B^2}} \, \mathrm{arth}\, \frac{\sqrt{1+B^2}\,\mathrm{th}\,(y/2)}{B + \mathrm{th}\,(y/2)} \cdot \frac{d\,(1/\sigma)}{d\Phi} +$$

$$+ \frac{1}{2} \cdot \frac{w_b^2}{D_p} K \frac{2kT}{q} \cdot \frac{b}{(1+b)^2} \cdot \frac{\mathrm{ch}\,y}{y^2\,(1+B^2)} \times$$

$$\times \left[\frac{2\,(y\,\mathrm{th}\,y - B^2)}{\sqrt{1+B^2}} \, \mathrm{arth}\, \frac{\sqrt{1+B^2}\,\mathrm{th}\,(y/2)}{B + \mathrm{th}\,(y/2)} + \right.$$

$$\left. + \frac{\mathrm{sh}\,(y1 + y\,\mathrm{th}\,y)\,(B\,\mathrm{th}\,y/2 - 1) + y\,(1+B^2)}{B + \mathrm{sh}\,y} \right] +$$

$$+ \frac{1}{2} \cdot \frac{w_b^2}{D_p} K \frac{kT}{q} \cdot \frac{b-1}{(b+1)} \cdot \frac{\mathrm{ch}\,y\,[y - \mathrm{th}\,y\,(1+y\,\mathrm{th}\,y)]}{y\,(B + \mathrm{sh}\,y)} \tag{173}$$

For convenience the term w_b/L is designated by y.

The transit time of carriers through the base, $t_{tr} = \frac{1}{2}w_b^2/D_p$, is equal to half the ratio of the square of the base thickness to the diffusion length of minority carriers. Therefore, by expanding B and assuming that the conductivity σ varies only as a result of changes in the concentration of equilibrium carriers, we obtain:

$$\frac{dU_b}{d\Phi} = \frac{4kT}{q} \frac{b}{(1+b)^2} \frac{\mathrm{ch}\,y}{\sqrt{1+B^2}} \, \mathrm{arth}\, \frac{\sqrt{1+B^2}\,\mathrm{th}\,(y/2)}{B + \mathrm{th}\,(y/2)} \frac{dn}{d\Phi} + \tag{173a}$$

$$+ t_{tr} K\,[F_1\,(y,\,B) + U'_D\,(y,\,B)],$$

where

$$F_1(y,\,B) = \frac{2kT}{q} \frac{b}{(1+b)^2} \frac{\mathrm{ch}\,y}{y^2\,(1+B^2)} \times$$

$$\times \left[\frac{2\,(y\,\mathrm{th}\,y - B^2)}{\sqrt{1+B^2}} \, \mathrm{arth}\, \frac{\sqrt{1+B^2}\,\mathrm{th}\,(y/2)}{B + \mathrm{th}\,(y/2)_,} + \right. \tag{173b}$$

$$+ \frac{\text{sh } y\,(1+y\text{ th }y)\,(B\text{ th }(y/2)-1)+y\,(1+B^2)}{B+\text{sh }y}\Bigg]$$

and

$$U'_D = \frac{kT}{q}\,\frac{b-1}{b+1}\,\frac{\text{ch }y\,[y-\text{th }y\,(1+y\text{ th }y)]}{y\,(B+\text{sh }y)}\,. \tag{173c}$$

The function F is the contribution (to the change in U_b) of the decrease in the depth of modulation of the base material conductivity caused by the radiation-induced decrease in the lifetime of minority carriers, while the function U'_D represents the contribution of the change in the Dember e.m.f.

Let us analyze the contribution of each term in equation (173a) to the total change in U_b. If the semiconductor is operated at high injection levels, i.e., with deep modulation of the base material conductivity, the main contribution comes from the second and third terms since the radiation sensitivity of minority carrier lifetime is greater than that of the concentration of majority carriers. However, the relative contribution of the three terms is determined not only by the injection level (parameter B) but also by the w_b/L ratio.

The dotted line on Figure 81 shows the dependence of F_i and the continuous line that of (F_i+U_D) on w_b/L, for four values of B. The monotonic increase of those functions with increasing w_b/L indicates, on one hand, that U_r depends on the base thickness (at constant L) and, on the other hand, that irradiation (i.e., a continuous increase of w_b/L,) causes a continuous increase in U_b. The changes in the first term (normalized to the product $t_r \cdot K$) for two values of B ($B = 0.1$ and $B = 10$) and bases of different thicknesses ($w_b = 0.5 \cdot 10^{-2}$ cm and $w_b = 2 \cdot 10^{-2}$ cm) are also shown in the figure. The conductivity σ and the rate of carrier removal $dn/d\Phi$ were assumed to be equal to 0.1 (ohm $\cdot$ cm)$^{-1}$ and 4.0 cm^{-1} respectively. The coefficient K for that value and for the corresponding injection level B was found from the Messenger equation (141) (see also Figure 57b). The above functions are shown in Figure 80 (1) and (2) ($B=0.1$) and (3) and (4) ($B=10$). The intersection points of curves corresponding to those functions (circles) with curves for (F_i+U_D) represent identical rates of change in U_b as a result of carrier removal (change in σ) and of the radiation-induced decrease in lifetime.

The absence of such intersection points (curve 3, $B = 10$ and

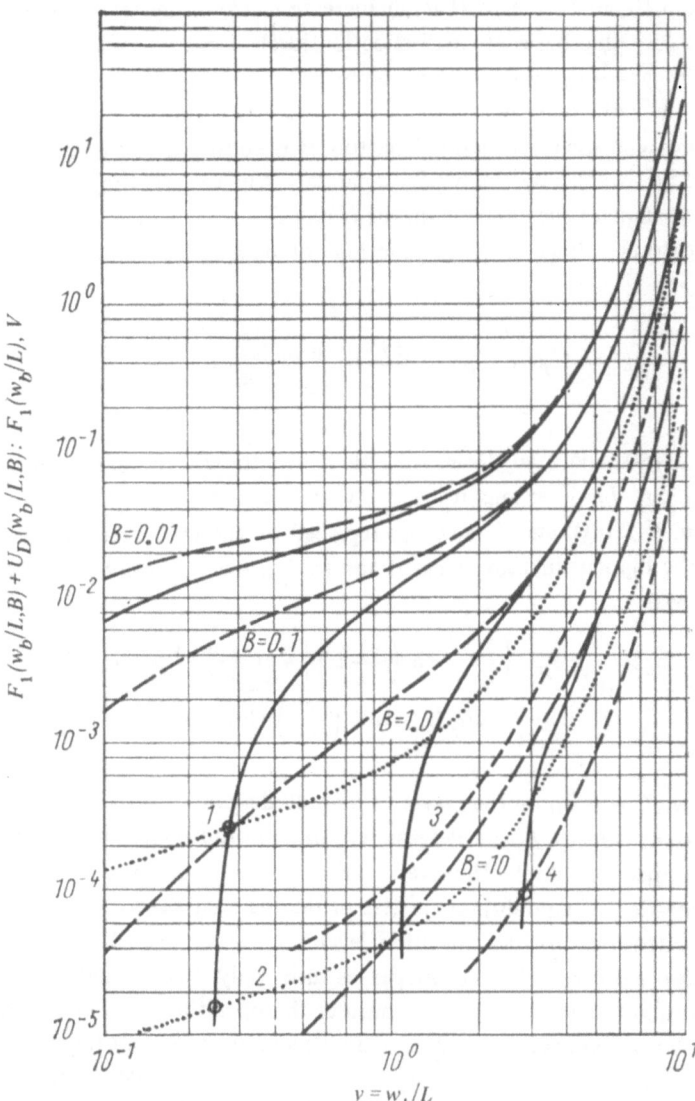

Figure 81. Dependence of the function F_i (dotted lines) and the sum of the functions $(F_i + U_D)$ (continuous lines) on the w_b/L ratio for various injection levels (various values of B). The numbers designate the dependence of the term in (173a) that determines the rate of change in U_b as a result of changes in the conductivity of the material. These curves were plotted for diodes with:

$w_b = 0.5 \cdot 10^{-2}$ cm and $B = 0.1$ (1); $w_b = 2 \cdot 10^{-2}$ cm and $B = 0.1$ (2); $w_b = 0.5 \cdot 10^{-2}$ cm and $B = 10$ (3); $w_b = 2 \cdot 10^{-2}$ cm and $B = 10$ (4) assuming that $\sigma = 0.1$ ohm·cm and $dn/d\Phi = 4.0$ (neutron·cm)$^{-1}$.

201

$w_b = 0.5 \cdot 10 \text{ cm}^{-2}$) means that an increase in U_b during irradiation of diodes with thin bases at small injection levels is caused exclusively by the change in σ and is not influenced by τ.

Let us now consider diodes with a p-i-n structure in which both junctions can inject carriers. Here, too, it is convenient to discuss the change in voltage by analyzing separately the voltage drops on the junctions and on the base, i.e. on the i-layer.

The total voltage on both junctions of p-i-n diodes can be found from:

$$U_{tr} = U_{pi} + U_{in} = \frac{2kT}{q} \ln \frac{\sqrt{\Delta p(0) \cdot \Delta p(w_i)}}{n_i}, \qquad (174)$$

where $\Delta p(0)$ and $\Delta p(w_i)$ are the hole concentrations in the i-region of the p-i and i-n junctions respectively and w_i is the thickness of the i-region. If $\Delta p(0)$ and $\Delta p(w_i)$ are replaced by the w_i/L_0 ratio (where L_0 is the bipolar diffusion length) and by J_{pr}, then U_{tr} can be written as follows:

$$U_{tr} = \frac{2kT}{q} \ln \frac{J_{pr} \left\{ \left[b + \text{ch}\left(\frac{w_i}{L_0}\right) \right] \left[b\,\text{ch}\left(\frac{w_i}{L_0}\right) + 1 \right] \right\}^{1/2}}{qD_0(b+1) n_i \left(\frac{w_i}{L_0}\right) \text{sh}\left(\frac{w_i}{L_0}\right)} \qquad (175)$$

The voltage drop on the i-layer of p-i-n diodes and its dependence on the current density J_{pr} are determined, like in ordinary diodes, by the carrier injection level. However, unlike ordinary diodes, the low concentration of equilibrium carriers in the material leads, even at low current densities, to high injection levels in p-i-n diodes which means that the existence of the i-layer is due only to the motion of non-equilibrium carriers. Let us consider the operation of p-i-n diodes with a linear dependence of J_{pr} on $\Delta p(0)$ and on $\Delta p(w_i)$ at the i-layer boundaries. In that case the voltage drop on the i-layer is not influenced by the current density J_{pr}. Both J_{pr} and the conductivity of the i-layer, modulated by the injected carriers, are linear functions of the hole and electron concentrations in the i-layer, provided that their mobility and their lifetime are constant and are not influenced by the concentrations.

A solution of the continuity equation for the voltage drop U_i on the i-layer yields

$$U_i = \frac{4kT}{q} \cdot \frac{b}{b+1} \cdot \frac{\operatorname{sh} \frac{w_i}{L_0}}{\left[b^2 + 2b \operatorname{ch} \frac{w_i}{L_0} + 1 \right]^{1/2}} \left\{ \operatorname{arctg} \left[\frac{b + e^{w_i/L_0}}{b + e^{-w_i/L_0}} \right]^{1/2} - \right.$$

$$\left. - \operatorname{arctg} e^{-w_i/L_0} \left[\frac{b + e^{w_i/L_0}}{b + e^{-w_i/L_0}} \right]^{1/2} \right\} + \frac{kT}{q} \cdot \frac{b-1}{b+1} \ln \frac{b \operatorname{ch} w_i/L_0 + 1}{b + \operatorname{ch} w_i/L_0} . \tag{176}$$

By differentiating equations (175) and (176) with respect to Φ we obtain the following expression for the forward voltage drop on *p-i-n* diodes:

$$\frac{dU_d}{d\Phi} = \frac{kT}{q} K\tau \frac{b}{b+1} \cdot \frac{w_i}{L_0} \left\{ \left[\frac{1}{2} \operatorname{sh} \frac{w_i}{L_0} \left(\frac{1}{b + \operatorname{ch} \frac{w_i}{L_0}} + \frac{b}{b \operatorname{ch} \frac{w_i}{L_0} + 1} \right) - \right. \right.$$

$$\left. - \operatorname{cth} \frac{w_i}{L_0} - \frac{1}{w_i/L_0} \right] + \tag{177}$$

$$+ \left[\frac{\operatorname{sh} \frac{w_i}{L_0}}{\left[b^2 + 2b \operatorname{ch} \frac{w_i}{L_0} + 1 \right]} \left(\frac{b \left(b + \operatorname{ch} \frac{w_i}{L_0} \right)}{b \operatorname{ch} \frac{w_i}{L_0} + 1} + \frac{b \operatorname{ch} \frac{w_i}{L_0} + 1}{b + \operatorname{ch} \frac{w_i}{L_0}} \right) + \right.$$

$$+ \frac{2 \left(b + \operatorname{ch} \frac{w_i}{L_0} \right) \left(b \operatorname{ch} \frac{w_i}{L_0} + 1 \right)}{\left[b^2 + 2b \operatorname{ch} \frac{w_i}{L_0} + 1 \right]^{3/2}} \left(\operatorname{arctg} e^{w_i/L_0} \left[\frac{b + e^{-\frac{w_i}{L_0}}}{b + e^{\frac{w_i}{L_0}}} \right]^{1/2} - \right.$$

$$\left. \left. \left. - \operatorname{arctg} \left[\frac{b + e^{-\frac{w_i}{L_0}}}{b + e^{\frac{w_i}{L_0}}} \right]^{1/2} \right) \right] + \frac{b-1}{2b} \cdot \frac{(b^2 - 1) \operatorname{sh} w_i/L_0}{\left(b + \operatorname{ch} \frac{w_i}{L_0} \right) \left(b \operatorname{ch} \frac{w_i}{L_0} + 1 \right)} \right\}$$

The first term in (177) is negative and represents the voltage changes on the *p-i* and the *i-n* junctions. The second term is always positive and is equal to the rate of change in the voltage drop on the *i*-layer. Figure 82 shows the dependence of those two terms and of their sum on w_i/L_0. At the intersection point of the $A_i dU_i/d\Phi$ and $A_i dU_{tr}/d\Phi$ curves, the rate of voltage change on the semiconductor is $dU_d/d\Phi = 0$.

The above equations represent qualitatively the behavior of irradiated diodes; the calculations have been confirmed experimentally. The forward V–I characteristics of two diodes, with ohmic back contacts, irradiated with different integrated fast neutron fluxes (Figure 83),

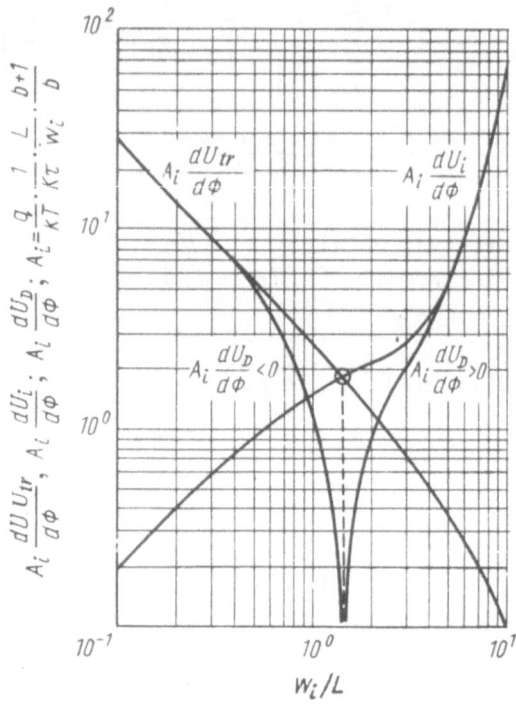

Figure 82. Dependence of the first two terms in equation (177) and of their sum on w_i/L_0 for p-i-n diodes.

show that at $B = 0.01$, when the currents through diodes with $w_b = 4.1 \cdot 10^{-2}$ and $w_b = 4.1 \cdot 10^{-2}$ cm were 1.35 and 5.35 A respectively, the forward voltage of the diode with a thicker base increased by a factor of 30.

The influence of w_b (or w_b/L) on the rate of change of the forward voltage is illustrated by Figure 84, which shows the characteristics of several diodes with bases of different thicknesses after irradiation with an integrated fast neutron flux of 10^{12} cm^{-2}.

Similar results have been obtained with other p-i-n diodes (Figure 85) /265/. Shwartz et al. /265/ found that those results agree both qualitatively and quantitatively with results calculated with the aid of equations (175) and (176).

The p-i-n diodes and diodes with an ohmic back contact are idealized mathematical models which do not exist in reality. Thus, in real p-i-n diodes the i-layer is never intrinsic, but only has a high resistivity and a

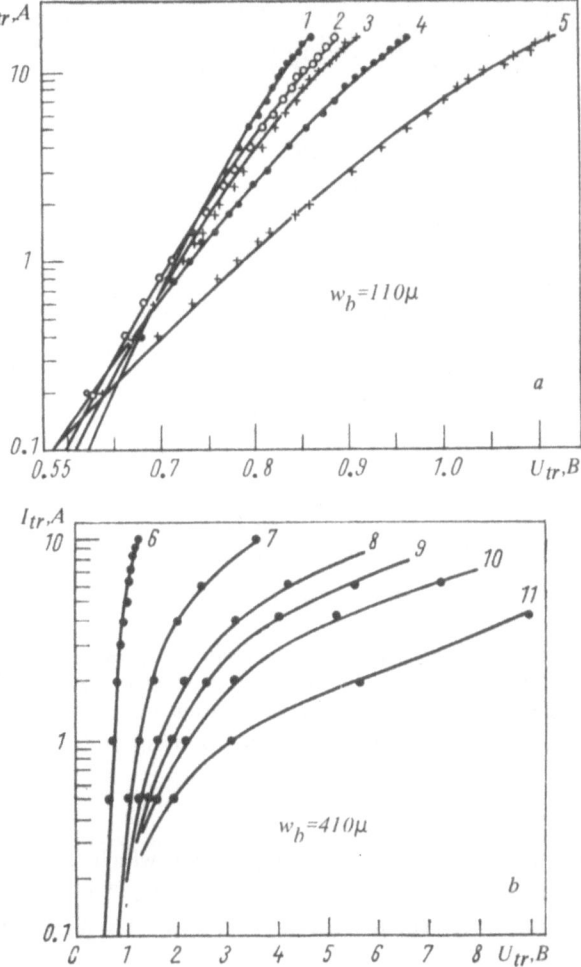

Figure 83. Forward V-I characteristics of two diodes with identical junction cross section areas A_{pn}=0.25 cm², prepared by diffusion of boron in silicon, after irradiation with various integrated fast neutron fluxes. Those diodes had an identical ρ=12.0 ohm·cm, but bases of different thicknesses [w_b=1.1·10⁻² cm (a) and w_b=4.1·10⁻² cm (b)].

1 - Φ = 0; 2 - Φ = 2.5·10¹² cm⁻²; 3 - Φ =5·10¹² cm⁻²; 4 - Φ = 2·10¹³cm⁻²; 5 - Φ=4·10¹³cm⁻²; 6 - Φ=0; 7 - Φ=2·10¹² cm⁻² 8 - Φ=9·10¹²cm⁻²; 9 - Φ = 6·10¹²cm⁻²; 10 - Φ = 10¹⁴ cm⁻²; 11 - Φ = 2.5·10¹³ cm⁻².

clearly pronounced *p*- or *n*-type conductivity. Similarly, a diode with a single junction never has a purely ohmic back contact. That contact

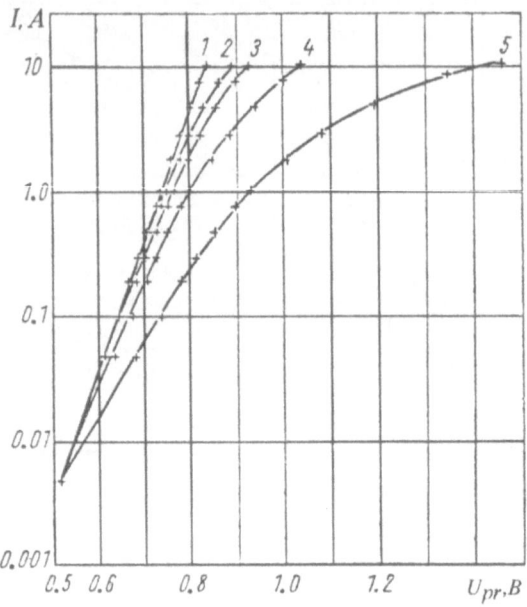

Figure 84. Forward V-I characteristics of diodes with bases of different thicknesses but with an identical junction surface area (A_{pn}=0.25 cm^2), irradiated with a fast neutron flux Φ=10^{12}cm^{-2}). 1 - 1.1·10^{-2} cm; 2 - 1.6·10^{-2} cm; 3 - 2.1·10^{-2} cm; 4 - 3.1·10^{-2} ; 5 - 4.1·10^{-2} cm.

may often be a $n+ -n$ (or $p+ -p$) junction which under certain conditions can inject carriers into the base region. The properties of such diodes are intermediate between those of the above idealized models.

Irradiation of silicon diodes, with resistivities of several ohm·cm, may cause a gradual change in their internal structure. Thus, diodes which at the beginning had properties similar to those of diodes with an ohmic contact, acquired a p-i-n structure. This is confirmed by the results in Figure 86, which shows the potential distribution in a diode with respect to the p-region, determined by the method of a moving probe. It is evident that an increase in fluence leads to increased voltage drop in the base region and in the ohmic contact, which starts to inject carriers. In low-resistance diodes the ohmic properties of the $n+ -n$ junctions are conserved over a wide range of radiation doses (Figure 87).

206

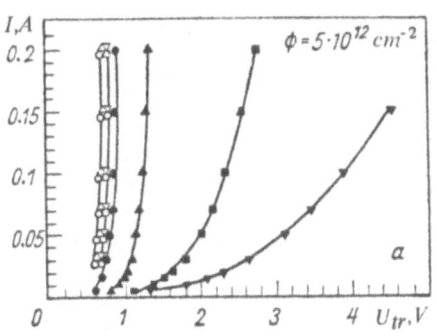

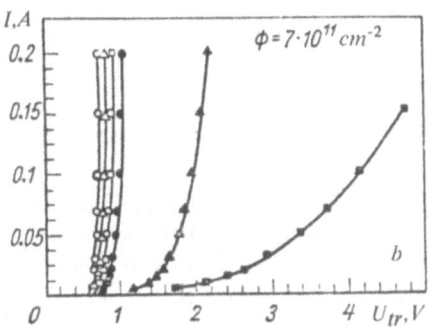

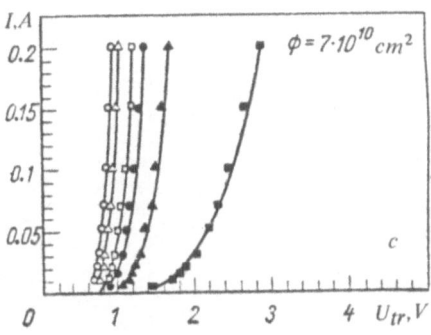

Figure 85. Forward V–I characteristics of *p-i-n* diodes with i-layers of different thicknesses before (○) and after (●) irradiation with fast neutrons.

a - w_b = 120 (○), 240 (▲) 360 (■) and 480 (▼) μ;

b - w_b = 480 (○), 720 (▲), 960 (■) μ;

c - w_b = 1200 (○), 1440 (▲), 1680 (■) μ.

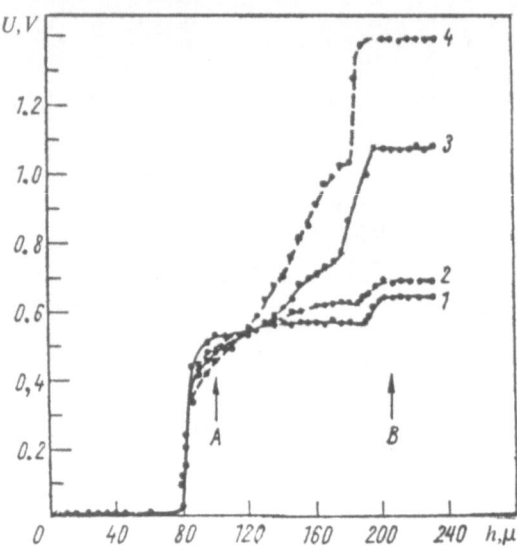

Figure 86. Distribution of potential in a forward-biased diode before and after irradiation as a function of fast-neutron flux: $1 - 0$; $2 - 10^{13}$ cm^{-2}; $3 - 4 \cdot 10^{13}$ cm^{-2}; $4 - 6 \cdot 10^{13}$ cm^{-2}; (AB - base thickness = $1.05 \cdot 10^{-2}$ cm; ρ_b = 32 ohm·cm; A_{pn} = 0.25 cm^2 I_{tr} = 0.75 A).

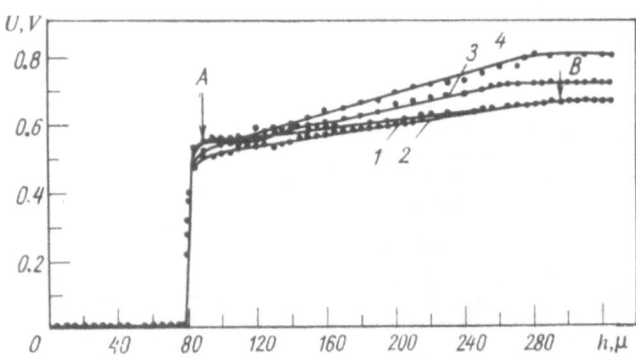

Figure 87. Potential distribution in a diode made of low-resistivity silicon before and after irradiation with fast neutrons. $1 - \Phi = 0$; I_{tr} = 0.75 A; $2 - \Phi = 6 \cdot 10^{13}$ cm^{-2}, I_{tr} = 0.56 A; $3 - \Phi = 6 \cdot 10^{13}$ cm^{-2}, I_{tr} = 0.75 A; $4 - \Phi = 2 \cdot 10^{14}$ cm^{-2}, I_{tr} = 0.75A (AB-base thickness = $2.05 \cdot 10^{-2}$ cm; ρ_b = 2.1 ohm·cm; A_{pn} = 0.25 cm^2.

2. Experimental Data on Changes in the Conductivity of Germanium and Silicon upon Irradiation.

Local levels in the forbidden band of semiconductors associated with radiation-induced defects and with the formation of defect clusters change the conductivity or the resistivity of the material by changing both the concentration of carriers and their mobility. However, noticeable changes in mobility occur, usually after the concentration of carriers has changed considerably as a result of capture by radiation-produced centers. Cleland *et al.* /217/ found that irradiation of *n*-type germanium (starting carrier concentration $2.8 \cdot 10^{16}$ cm^{-3}) with a $2.9 \cdot 10^{15}$ cm^{-2} fast neutron fluence at room temperature, reduced the mobility of electrons by only 20%. However, this decrease in mobility must be taken into account in the calculation of the conductivity of irradiated semiconductors at low temperatures when scattering on the charged centers starts to prevail.

Our analysis of the influence of radiation on the conductivity of semiconductors at room temperature is based on the assumption that it is determined solely by changes in the carrier concentration.

The centers (produced in germanium by radiation) capture carriers or are depleted of carriers so that the concentration of electrons in the conduction band changes monotonically, approaching a certain maximum value which is smaller than the intrinsic carriers concentration. Therefore, the conductivity of *n*-type germanium at room temperature, regardless of its initial value, continuously decreases during irradiation and finally reaches a value near the intrinsic conductivity of the material. Further irradiation of *n*-type germanium causes a conversion, as a result of which the material acquires *p*-type conductivity which increases asymptotically and approaches a certain limiting value σ_{lim} (Figure 88) /266/. The *p*-type conductivity decreases if its initial value exceeds the limit and, on the contrary, it increases if its starting value is smaller than the limit.

The conductivity of *p*- or *n*-type silicon decreases monotonically during irradiation to a value close to its intrinsic conductivity, (Figure 89) /266/.

The carriers concentration in irradiated semiconductors can be found by determining the distribution of electrons or holes between impurities introduced by doping and radiation-induced defects some of which may be polyvalent. The solution of this problem is described in /267/. In some cases it may be solved by a graphic presentation of the

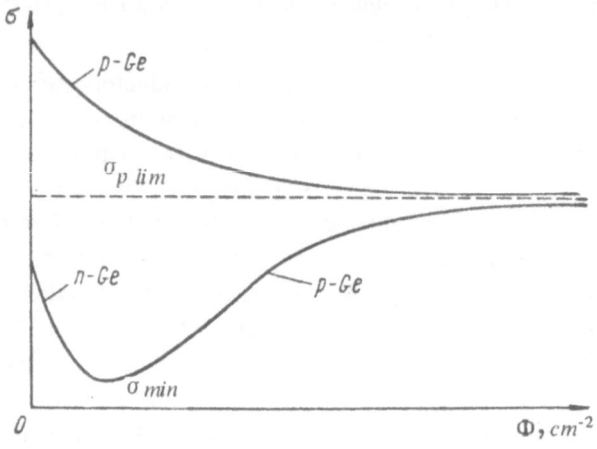

Figure 88. Changes in the conductivity of p-type and n-type germanium during irradiation with fast neutrons.

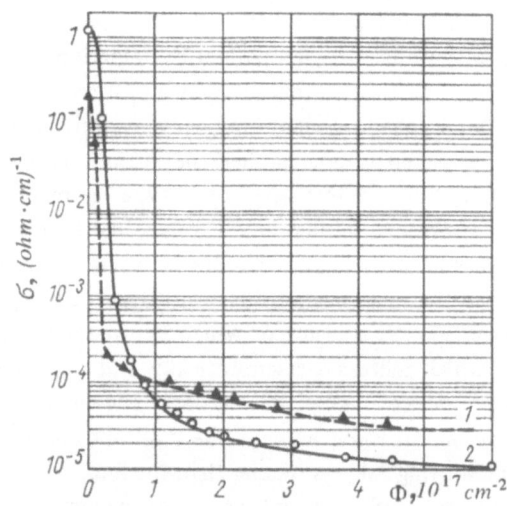

Figure 89. Changes in the conductivity of p-type and n-type silicon irradiated with fast neutrons. 1-p-Si(63°C):2-n-Si(48°C).

dependence of the Fermi level position on the concentration of deep-lying donor and acceptor levels, as suggested by James *et al.* /268/.

The concentration of electrons or holes in n-type semiconductors at equilibrium thermal conditions can be found by solving the quadratic equation relative to the intrinsic concentration n_i and to the difference between the electron and hole concentrations $N_r = n\text{-}p$;

$$n = \frac{1}{2} N_r \left[\left(1 + \frac{4n_i^2}{N_r^2} \right)^{\frac{1}{2}} + 1 \right] ;$$

(178a)

$$p = \frac{1}{2} N_r \left[\left(1 + \frac{4n_i^2}{N_r^2} \right)^{\frac{1}{2}} - 1 \right] .$$

(178b)

When $N_r \gg n_i$ the equations are simplified:

$$n \approx N_r + n_i^2 / N_r,$$

(179a)

$$p \approx n_i^2 / N_r.$$

(179b)

The concentration differences can be found from the condition of preservation of neutrality. If we write N_d and N_a for the concentrations of chemical donors and acceptors and N_{ta} and N_{td} for the concentrations of radiation induced defects with acceptor and donor properties respectively, then the condition of neutrality can be written as follows:

$$N_r = N_d - N_a + \frac{N_{td}}{1 + \beta_d^{-1} \exp\left(\frac{E_a - E_F}{kT} \right)} -$$

(180)

$$- \frac{N_{ta}}{1 + \beta_a \exp\left(\frac{E_a - E_F}{kT} \right)} ,$$

where E_F is the Fermi level; E_d and E_a are the donor and acceptor levels of radiation defects and β_d and β_a are the spin degeneracy factors for those donor and acceptor levels.

Let us consider two cases in which the defect levels in a n-type semiconductor are such that in one case $(E_F - E_d) > 0$ and $(E_F - E_a) > 0$ while in the second case $(E_F - E_d) > 0$ while $(E_F - E_a) < 0$. In both cases the defect levels are completely filled and therefore they have no effect on the distribution of electrons.

From (180) it follows for the first case:

$$N_r = N_d - \dot{N}_a - N_{ta},$$

(181)

$$\frac{dN_r}{d\Phi} = -\frac{dN_{ta}}{d\Phi} \quad \text{hence,}$$

$$\frac{dn}{d\Phi} = -\frac{dN_{ta}}{d\Phi} \left(1 - \frac{n_i^2}{N_r^2}\right) \approx -\frac{dN_{ta}}{d\Phi},$$

i.e., the rate of carrier removal is almost equal to the rate of production of acceptor levels while the concentration of electrons decreases linearly with increasing fluence.

$$n = n_0 - \frac{dN_{ta}}{d\Phi} \Phi. \tag{182}$$

In the second case:

$$N_r = N_d - N_a - \left(\frac{1}{\beta_a}\right) N_{ta} \exp\left(\frac{E_F - E_a}{kT}\right) \tag{183}$$

and:

$$\frac{dN_r}{d\Phi} \approx \frac{1}{\beta_a} \cdot \frac{dN_{ta}}{d\Phi} \exp\left(\frac{E_F - E_a}{kT}\right) = -\frac{1}{\beta_a} \cdot \frac{n}{n_1} \frac{dN_{ta}}{d\Phi}.$$

Assuming that:

$$\frac{dn}{d\Phi} \approx \frac{dN_r}{d\Phi},$$

we obtain:
$$\frac{dn}{d\Phi} \approx -\frac{1}{\beta_a} \cdot \frac{n}{n_1} \cdot \frac{dN_{ta}}{d\Phi}, \tag{184}$$

where n_1 is the electrons concentration in the conduction band when the Fermi level coincides with the position of acceptor levels of defects.
From (184) we obtain:

$$n = n_0 \exp\left[-\frac{1}{\beta_a} \cdot \frac{1}{n_1} \frac{dN_{ta}}{d\Phi} \Phi\right], \tag{185}$$

i.e., the electron concentration decreases exponentially with increasing fluence.

Equations (182) and (185) describe only approximately the real changes in conductivity since in reality irradiation, particularly with neutrons, produces structural defects of various kinds. It has been found that the nature and properties of such defects are determined by the type of the semiconductor material and the type of radiation, by the amount and type of the major dopant not controlled by other

212

impurities and by the dislocation density. Each of the new defects may be characterized by the rate of production per unit of radiation, by its properties (acceptor or donor) and by the position of levels in the forbidden gap. All this makes difficult the quantitative determination of phenomena observed over a wide rsnge of integrated fluxes or radiation doses. However, the fact that the rate of production of certain defects (e.g. A-centers in oxygen-containing silicon) may be much higher than the rate of production of other defects, allows us to use the above equations to evaluate the radiation stability of semiconductor devices with an accuracy sufficient for practical purposes.

In relatively "pure" germanium the levels of radiation-induced defects are in the middle of the forbidden band. According to many authors one of those levels is 0.2(0.23) eV below the bottom of the conduction band. Its concentration n is close to $3 \cdot 10^{15}$ cm^{-3} which corresponds to a conductivity of about 0.5 ohm· cm. Konopleva and Novikova /269/ found that the conduction band also contains vacancy levels 0.3 eV above the valence band. The concentration n_1 of these levels is smaller than the intrinsic carrier concentration in germanium. Those two types of levels are produced at about the same rates.

The rate at which electrons would be removed from n-type germanium, (in the resistivity range from 2.0 $(ohm \cdot cm)^{-1}$ to the intrinisic) will be determined by the probability of filling of those levels, which differs from unity. Consequently, n-type germanium can be represented by a term similar to that given in (185) but obtained by solving a different equation which takes into account all produced defects. Therefore, the analytical presentation of $\sigma = F(\varphi)$ is difficult.

It is believed /269, 270/ that certain impurities such as Cu and Au, which are inactive in interstices, fill the vacancies and become electrically active as a result of irradiation. If those impurities produce acceptor levels in the lower half of the forbidden band, as is the case of copper, the rate of carrier removal in their presence would decrease. This assumption has been qualitatively confirmed by the data of Konopleva and Novikova /269/ (Table 12) which also indicate that the dislocation density in germanium has no effect on the rate of electron removal or hole production. In n-type silicon, produced by the Czochralski method and containing large amounts of oxygen (concen-

Table 12. Rate of carrier removal as a result of neutron irradiation of germanium and silicon of various dislocation densities and copper contents.

Type of material	Resistivity ohm·cm	Dislocation density cm^{-2}	Concentration of copper atoms cm^{-3}	Rate of carrier production or removal per single fast neutron cm^{-1}
n-Ge	1.0	Ordinary	—	—3.3
n-Ge	2.0	»	$\sim 5 \cdot 10^{14}$	—8.0
n-Ge	1.6	$\sim 10^2$	—	—3.6
n-Ge	5.0	$\sim 10^6$	—	—3.2
p-Ge	1.0	Ordinary	—	+0.13
p-Ge	7.0	$8 \cdot 10^2$	—	+0.15
p-Ge	33.0	$\sim 10^6$	—	+0.13
n-Si	1.0	Ordinary	—	—2.0
n-Si	1.0	»	$\sim 5 \cdot 10^{14}$	—1.4
p-Si	4.5	»	—	—1.5
n-Si	5—10	»	$\sim 5 \cdot 10^{14}$	—1.5
n-Si	8.0	$\sim 10^6$	—	—10^{-3}
n-Si	150.0	$\sim 10^5$	—	—10^{-3}
p-Si	$3 \cdot 10^3$	$\sim 10^5$	—	—10^{-3}

tration higher than 10^{17} cm^{-3}), the rate of production of A-centers is much higher than the rate of production of other centers.

According to Wertheim /232/ in silicon with a conductivity $\sigma \sim 0.1$ (ohm· cm)$^{-1}$, irradiated with 1 MeV electrons the rate of production of A centers ($E_c - 1.16$ eV) is 0.18 cm^{-1} while the rate of production of E centers ($E_c - 0.4$ eV) is only $5 \cdot 10^{-3}$ cm^{-3}. Hence, equation (185) can only be used for the calculation of the production of A centers in n-type silicon with a resistivity from a fraction of one to several tens of ohms · cm. In oxygen-free silicon and in higher resistivity silicon the acceptor levels near the middle of the forbidden band, and in particular those of E-centers, have a predominant effect on the change in conductivity as a result of irradiation.

Manlief/264/, on the basis of unpublished data of Stein and Longo, reported that the above is true for p-type silicon within the same range of resistivities.

The change in the effectiveness of carrier capture by the newly introduced levels determines the dependence of the rate of carrier removal on the position of the Fermi level. Such a dependence has been found by many authors/219,237,238/.

Kantz/219/ found that during irradiation of n-type silicon (obtained by zone-melting in vacuum) with 1.5 MeV neutrons the rate of removal of electrons per neutron with an energy > 1 keV increased with increasing electron concentration. The above rate is approximately equal to 2.0 cm^{-1} for silicon with $\rho = 100$ ohm· cm, 4.0 cm^{-1} for silicon with $\rho = 10$ ohm · cm and 8.5 cm^{-1} for silicon with $\rho = 1$ ohm · cm.

Konopleva and Novikov/269/ found that the rate of carrier removal from silicon is greatly influenced by the dislocation density. Their data (Table 12) indicate that the carrier removal from silicon with a high dislocation density $(10^5 - 10^6$ cm$^{-2})$ is much slower $(10^{-3}$ cm$^{-1})$ than the removal of carriers from ordinary commercial silicon regardless of the type of conductivity and resistivity of the material; this was attributed to the fact that the binding energy of interstitials and their clusters is higher $(2-3$ eV$)$ than the binding energy of vacancies $(0.03$ eV$)$. Therefore, many levels which are associated with interstitials cannot be formed in a material with a high dislocation density because the interstitials are attached to the dislocations.

The above numerical values of the rates of carrier removal and the rate of change of other semiconductor properties (as a result of irradiation) must be considered as approximate since the absolute values are determined not only by the properties of the material and by the type of radiation but also by the energy spectrum of the radiation. This was emphasized by the data of Kantz (Figure 90), who found that carrier removal from silicon per neutron with an energy >1 keV is proportional to the mean energy $\bar{E}$ in the spectrum range of 0.3-2.4 MeV. Obviously, other semiconductor properties, among them the lifetime of non-equilibrium carriers, could be similarly affected by the radiation spectrum.

Kantz determined the mean energy of neutrons in the spectrum using two threshold indicators (one was Pu^{239} with a shield of enriched B^{10} and the other was S^{32}). The boron shield was designed so that plutonium recorded the number of neutrons with an energy >1 keV. The threshold energy of the sulfur indicator was close to 3.0 MeV/271/.

Since dosimetry of fast neutrons with sulphur indicators is widely used it would be interesting to determine the energy dependence of the rate of defect production if the number of defects is normalized against

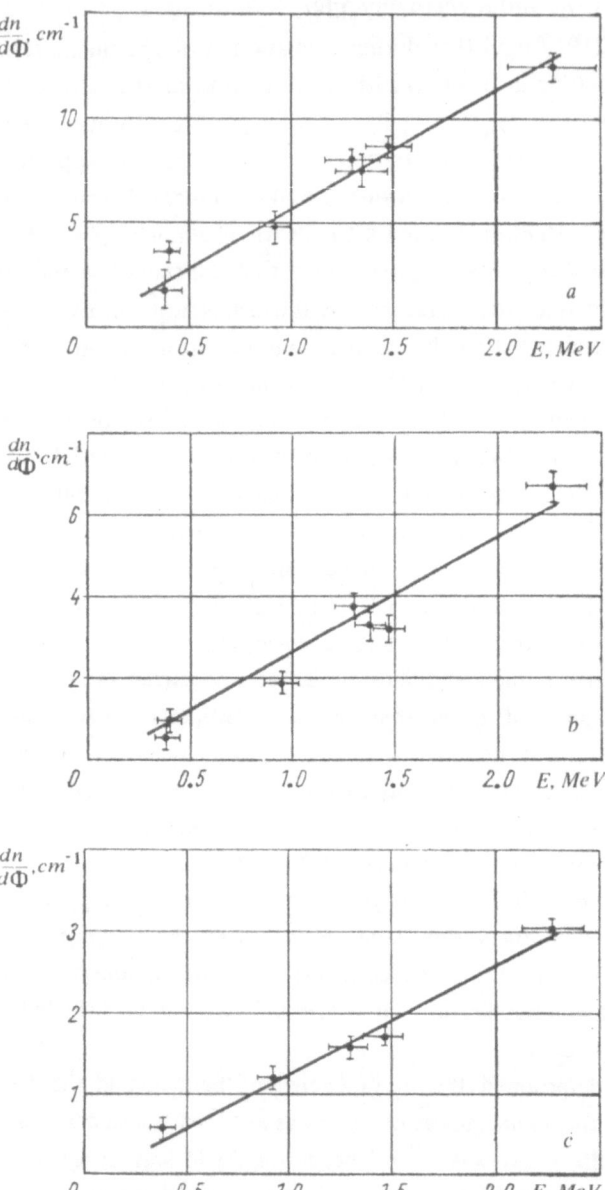

Figure 90. Dependence of the rate of carrier removal per $>1\mathrm{keV}$ neutron, on the mean neutrons energy E for oxygen-free n-type silicon samples of different resistivities /219/: a - 1.0; b - 10.0; c - 100 ohm.cm. Experimental points correspond to irradiation with reactor neutrons.

one neutron with an energy higher than the threshold energy of the sulphur indicator.

Figure 91 shows such functions for silicon samples of different resistivities, plotted on the basis of data obtained by Kantz. The number of removed electrons per neutron with an energy exceeding 3.0 MeV was plotted on the ordinate. The functions contradict those found by Kantz; the rate of defect production increases hyperbolically with the decrease in the mean energy of neutrons E. Such an energy dependence of the rate of defect production is due to the fact that in reality the radiation defects are produced not by neutrons with an energy exceeding 3 MeV but by neutrons with a lower energy. Therefore the softer the neutron spectrum (i.e., the lower the mean energy of neutrons in the spectrum) the larger is the number of defects per neutron with an energy exceeding the threshold energy of the sulphur indicator.

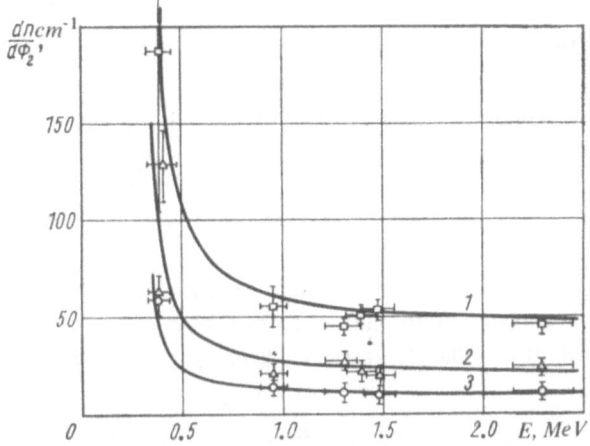

Figure 91. Effect of neutron energy on the rate of carrier removal per neutron ($>$ 3 MeV) in oxygen-free n-type Si with resistivities of 1, 10 and 100 ohm.cm (curves 1, 2, 3) according to /219/.

It should be pointed out that the rate of defect production remains fairly constant ($\pm15\%$) within the range of mean energies E from about 0.8 to 2.5 MeV, which encompasses various spectra. Therefore it is possible to compare results obtained under different conditions, provided that in all cases the damage is related to an integrated flux of neutrons with an energy $>$ 3.0 MeV.

3. Radiation–Induced Changes in the Reverse V–I Characterisitics of Diodes

The breakdown voltage is the most important parameter of the reverse V–I characteristics of diodes. Before breakdown, the most important characteristic is the reverse current. Those two parameters are influenced by irradiation.

The reverse current of real diodes comprises several components of different nature and with different dependences on the temperature and the applied voltage. Most important among these components are: the diffusion component which depends on the thermal production of minority carriers in the bulk of the semiconductor material, which is limited by the diffusion length from the p-n junction plane and a component associated with the production of carriers of both types in the space charge layer of the p-n junction. There are also components associated with possible leakage over the surface and with the formation of channels and inverse layers on the surface. In some cases the contribution of these components to the total reverse current is predominant.

The diffusion component of the reverse current is equal to the product of the rate of production of minority carriers and the effective volume from which they are picked up by the p-n junction.

$$I_{rev.d} = V_{eff}\frac{p_n}{\tau_{p.}} + V''_{eff}\frac{n_p}{\tau_n} \tag{186}$$

In "elongated" diodes which meet the condition $w_b \gg L$, the effective volume is determined by the surface area of the p-n junction and by the diffusion length of minority carriers. Therefore for such diodes, we can write:

$$I_{rev.d} = q\,\frac{D_p}{L_p}\,A_{pn}\,\frac{n_i^2}{n_n} + q\,\frac{D_n}{L_n}\,A_{pn}\,\frac{n_i^2}{p_p} \tag{187}$$

In real diodes the effective volume is usually limited either by the thickness of the base w_b or by the length of the diffusion shift over the surface, which depends on the rate of surface recombination $\sqrt{Dw_b/2s}$ (s is the rate of surface recombination), rather than by the diffusion length of minority carriers. Obviously, the diffusion component of the reverse current in diodes with n-type base can be expressed as follows:

$$I_{rev.d} = q \frac{w_b A_{pn}}{\tau_p} \cdot \frac{n_i^2}{n_n} \qquad (188a)$$

or

$$I_{rev.d} = q \pi D_p \frac{w_b^2}{2s} \cdot \frac{1}{\tau_p} \cdot \frac{n_i^2}{n_n} . \qquad (188b)$$

According to Sah *et al.*, the component generated in the *p-n* junction would be

$$I_{rev.pn} = \frac{q w_{pn} A_{pn} n_i}{2\tau_{pn} \text{ch} \left[(E_t - E_i) \Big/ kT + \frac{1}{2} \ln \frac{\tau_{p0}}{\tau_{n0}} \right]} \qquad (189)$$

where

$$\tau_{pn} = \sqrt{\tau_{p0} \tau_{n0}}$$

This component, unlike the diffusion component, is influenced by the reverse voltage on the diode since it determines the width w_{pn} of the space charge region

$$w_{pn} = \sqrt{\frac{2\varepsilon_0 \varepsilon (\psi_c + U_{rev})}{q n_n}} \qquad (190a)$$

for an abrupt *p-n* junction, and

$$w_{pn} = \sqrt[3]{\frac{12\varepsilon_0 \varepsilon (\psi_c + U_{rev})}{q_a}} \qquad (190b)$$

for a graded *p-n* junction with a dopant concentration gradient $a = dN_a/dx$.

At room temperature the diffusion component of the reverse current is predominant in germanium diodes and the component generated in the *p-n* junctions is predominant in silicon diodes.

The two components of the reverse current include parameters that are influenced by radiation, such as the lifetime of minority carriers and the concentration of majority carriers. The changes in those parameters upon irradiation always cause an increase in the reverse current. By substituting the terms for the radiation dependence of the lifetime of minority carriers and for the concentration of majority carriers in equation (188a) we obtain:

219

$$\frac{dI_{rev.d}}{d\Phi} = qA_{pn}w_b\frac{n_i^2}{n_n} K +$$

$$+ q\frac{A_{pn}w_b}{\tau_p} \cdot \frac{n_i^2}{n_{n0}} \cdot \frac{1}{\beta_A} \cdot \frac{1}{n_1}\frac{dN_t}{d\Phi} e^{\frac{1}{\beta_A n_1} \cdot \frac{dN_t}{d\Phi}}\Phi \tag{191}$$

The first term, which is responsible for the linear dependence of I_{rev} (reverse current) on the integrated radiation flux is predominant during the first stage of irradiation, since the process is too fast to change the concentration n_n. Further irradiation leads to an exponential increase in the current with increasing fluence. Such a dependence of I_{rev} on Φ has been indeed found in germanium diodes .in which the diffusion component is predominant. Figure 92 shows experimental data on the dependence of I_{rev} on Φ for a low-power germanium diode; the data agree with the theoretical predictions.

In silicon diodes, in which the p-n junction component is predominant, the reverse current would be expected to increase linearly with increasing Φ over a wide range of fluences since Φ is more

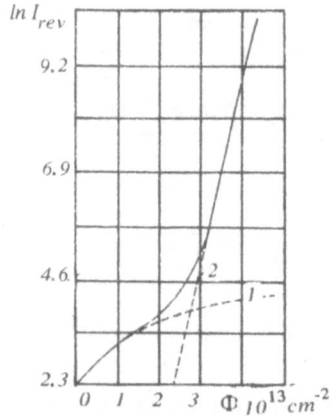

Figure 92. Changes in the current in alloy-type germanium diodes irradiated with fast neutrons. The experimental dependence (continuous line) can be represented by the superposition of two functions of the integrated flux Φ : A linear function (1), I'_{rev} $\Phi = I_{rev.0} + N\Phi$ and an exponential function (2) $I''_{rev} = P + Qe\lambda\Phi$

closely associated with the lifetime of carriers than with the conductivity of the material or with the concentration of majority carriers. Such a relationship has indeed been dound to exist; however, it is often difficult to distinguish it from the basic components of the leakage current.

The breakdown voltage is associated with the concentration of majority carriers, both before and during irradiation. If irradiation reduces the conductivity of the material, the breakdown voltage can be expected to increase during both Zener breakdown and avalanche multiplication of carriers.

Zener breakdown takes place only in thin p-n junctions and therefore it develops only in diodes made of highly doped low resistivity materials.

According to Ya. A. Fedotov/272/ such a breakdown in germanium takes place at resistivities lower than 0.1-0.2 ohm·cm. For such junctions the Zener breakdown voltage can be related to the material characteristics by the following equation /194/:

$$U_{br} = \frac{\varepsilon_0 \varepsilon E_{cr}^2 \mu_b}{2} \rho_b, \tag{192}$$

where E_{cr} is the critical field strength.

Therefore, the Zener breakdown voltage increases when the conductivity of irradiated n-type germanium and n-type or p-type silicon decreases as a result of a decrease in the number of free carriers.

In high-voltage silicon diodes, with a breakdown voltage above 100 V, breakdown takes place as a result of avalanche multiplication of carriers in the p-n junctions. Since the carriers must acquire an energy sufficient to produce an electron-hole pair, over a path equal to the free length, the critical field strength at the junction, at which an avalanche starts, depends on the width of the space charge layer. The value of E_{cr} decreases with increasing junction width (increasing resistivity of the semiconductor). Therefore, the effect of the resistivity or conductivity on the breakdown voltage is smaller than that of Zener breakdown. It has been found that $U_b = B\rho_b^k$ where $B = 52$ and $k = 0.61$ for germanium junctions of the n^+-p type, $B = 83.4$ and $k = 0.6$ for germanium junctions of the p^+-n type, $B = 23$ and $k = 0.75$ for silicon junctions of the n^+-p type and $B = 86$ and $k = 0.64$ for silicon junctions of the p^+-n type.

For silicon junctions the breakdown voltage can be calculated from an empirical equation:

$$U_b = 8.65 \cdot 10^{11}/a^{0.508}.$$

Since irradiation increases the resistivity of silicon, the avalanche breakdown voltage for abrupt junctions can also be expected to increase but that increase is smaller than in the case of Zener breakdown. The same is true for graded junctions, although this is not obvious.

We mentioned above that the conductivity of irradiated silicon decreases monotonically and asymptotically approaches a value near the intrinsic conductivity, regardless of the conductivity type. The rate of carrier removal increases with increasing equilibrium carriers concentration. In graded junctions the concentration of carriers increases more rapidly as the distance from the center of the junction is increased. As a result, the carriers concnetration gradient in such a junction decreases monotonically while the breakdown voltage increases. The increase in voltage across such junctions is less pronounced than in abrupt junctions.

The above analysis of the influence of radiation on the breakdown voltage of junctions has been confirmed experimentally. Thus, Figure 93 shows that the increase in the breakdown voltage of silicon junctions, (produced by diffusion of aluminum in n-type silicon with resistivities of 2 and 20 ohm·cm) caused by an increase in the fast neutron fluence is greater than in the case of junctions in materials with a higher resistivity. In addition, the characteristics of the prebreakdown region become "soft" and the reverse current begins to increase noticeably, before a stable breakdown takes place; this is apparently casued by the formation of heterogeneities in the p-n junction.

4. Effect of Radiation on the Parameters of "Stabilitrons."

Silicon "stabilitrons" (which are analogous to gas-discharge stabilizers used in semiconductor electronics) are diodes operating in the breakdown range of the reverse V—I characteristics. Stabilitrons are used chiefly for voltage stabilization, as sources of a reference voltage and other similar purposes, and hence their name.

Naturally, their main parameters are those of the reverse V—I characteristics such as the breakdown voltage, which in this case is called stabilization voltage (U_s), its dependence on the working breakdown voltage (dynamic resistance during stabilization) and the temperature coefficient of the stabilization voltage (TCV).

In addition, since stabilitrons are ordinary diodes with alloyed or diffused p-n junctions, they have excellent characteristics in the range

222

of forward V–I characteristics and therefore can be used as ordinary low-voltage diodes with a high radiation stability.

The *p-n* junction in stabilitrons may exhibit Zener breakdown and an avalanche breakdown or combined breakdown, depending on the resistivity of the material. In diodes with a low-resistivity base and consequently, with a low breakdown voltage, Zener breakdown is most probable. Avalanche breakdown would develop in diodes with high-resistivity bases. The prevailing breakdown mechanism can be determined fromt he sign of the TCV. In diodes with Zener breakdown the dependence of the breakdown voltage on the forbidden band width, (which becomes narrower as the temperature increases) is a linear function. The narrowing of the forbidden band reduces both the critical field voltage E_{cr} and the breakdown voltage. Therefore TCV becomes negative, which means that the breakdown voltage decreases with increasing temperature.

In diodes subject to avalanche breakdown, the breakdown voltage is inversely proportional to the mobility of carriers, since it is proportional to the resistivity of the diode. Since in the working range of temperatures (-60 to $+120°C$) the mobility of carriers decreases with increasing temperature, TCV should be positive. The limiting breakdown voltage at which TCV changes its sign, and therefore is close to zero, is equal to about 5.5 V. Mass produced Soviet stabilitrons have a working voltage above 6–7 V and therefore their TCV is always positive $(0.05-0.1\%/°C)$ /273/. In order to reduce the TCV to a level not exceeding that of normal components, several diodes can be connected in series and one of them is operated as a stabilitron in the breakdown region of the reverse characteristics while the others are operated in the forward region with a negative TCV.

The influence of radiation on the stabilization voltage is of great interest for the study of the characteristics of irradiated stabilitrons. An earlier qualitative study has shown that the breakdown voltage increases monotonously during both avalanche and Zener breakdowns. However, this study requires quantitative confirmation.

Let us now make a quantitative analysis of low-voltage ($U_s \approx 7-10V$) alloyed stabilitrons. Assuming that in the investigated range of integrated fast neutron fluxes (smaller than $10^{15} cm^{-2}$) the critical field strength E_{cr} does not change much, and we can relate changes in U_s to the width of the space charge layer w_{pn}, we obtain from (109):

$$\frac{dU_{st}}{d\Phi} \Big/ U_{st} = - \frac{\dfrac{dn_n}{d\Phi}}{2n_n}, \qquad (193)$$

i.e., the relative rate of change in U_s is directly proportional to the relative rate of removal of electrons from the conduction band. Such

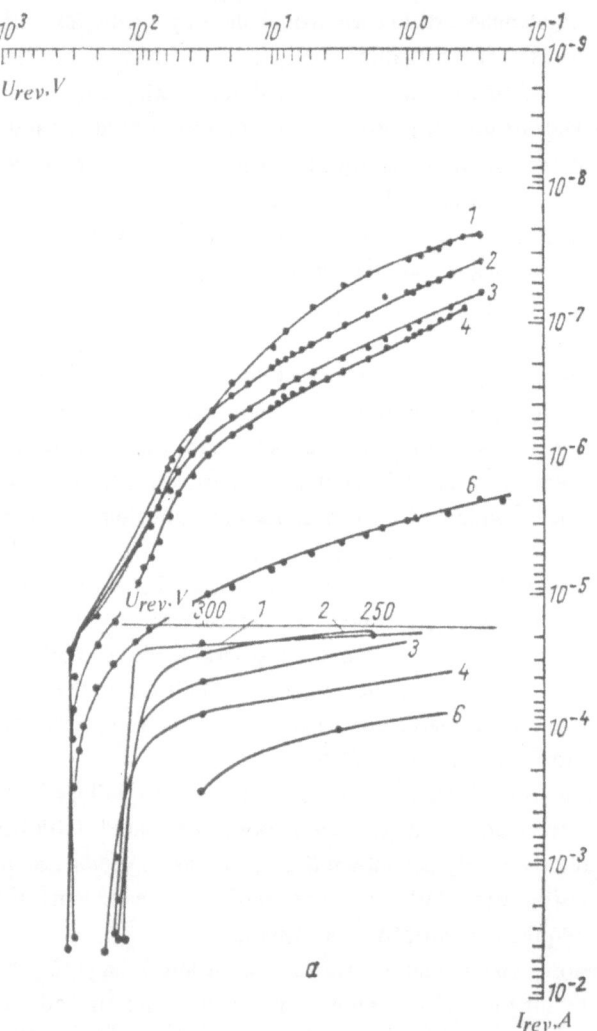

Figure 93. Influence of bombardment with fast neutrons on the breakdown voltage of diffused silicon junctions obtained by

$1 - \Phi = 0$; $2 - \Phi = 2.1 \cdot 10^{13} \text{cm}^{-2}$: $3 - \Phi = 4.2 \cdot 10^{13} \text{cm}^{-2}$; $4 - \Phi = 6.3 \cdot 10^{13} \text{cm}^{-2}$;

(continued on p. 225)

low-voltage stabilitrons are made of silicon with n-type conductivity (10 to 15 (ohm·cm)$^{-1}$) and with a carrier concentration from 10^{17} to 10^{18} cm^{-3} respectively. In such silicon the rate of carrier removal per single neutron is nearly constant and equal to the sum of the rate of production of A and E centers and of other acceptor levels produced by

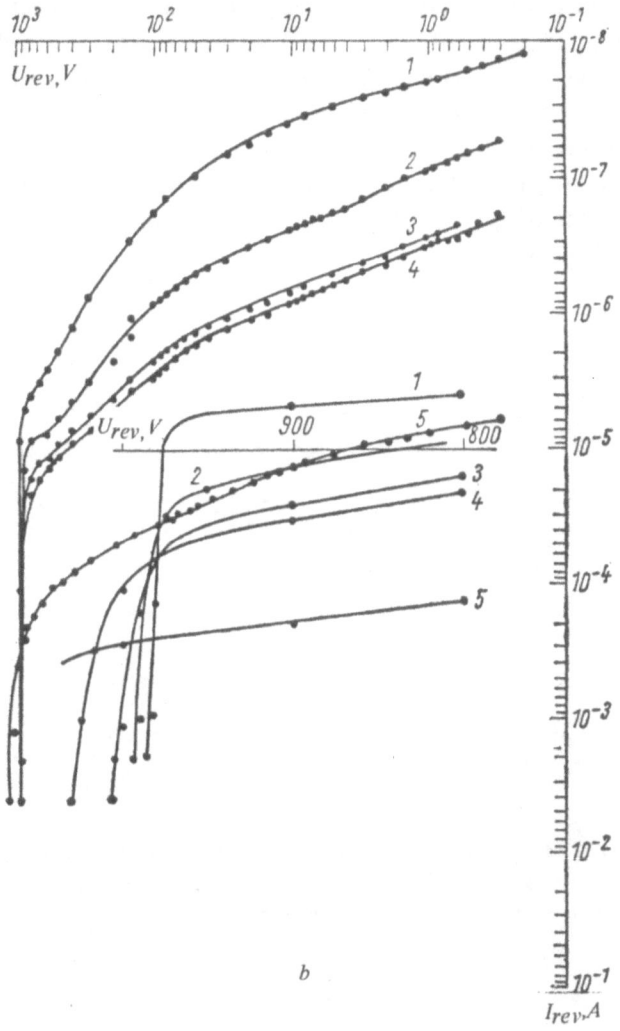

Figure 93 cont.
diffusion of boron in n-type silicon with $\rho=2$ ohm·cm (*a*) and $\rho=20$ ohm·cm (*b*):
$5 - \Phi = 3.8\cdot10^{14}$ cm^{-2}; $6 - \Phi = 4.4\cdot10^{14}$ cm^{-2}.

radiation. No experimental data are available on the rate of carrier removal from such silicon particularly for the neutron spectrum in which the stabilitrons were irradiated. However, using the data of Kantz (Figure 90) it is possible to assess the order of magnitude of that rate. Its minimum value is not less than 7.5 cm^{-1} and its maximum value does not exceed 20 cm^{-1}. Consequently the relative rate of carrier removal and the relative rate of change of the stabilization voltage must be within the range of $(0.4\text{-}1) \cdot 10^{-17}$ (neutron/cm^2)$^{-1}$! for high voltage stabilitrons of this group. The experimentally found values of $dU_s/d\Phi$ for stabilitrons are in the range of $(0.65\text{-}2.3) \cdot 10^{-17}$ (neutron/cm^2)$^{-1}$.

Experimental data on the dependence of $dU_s/d\Phi/U_s$ on the irradiation time in a reactor, indicate that at integrated fluxes up to 10^{15} neutron/cm^2 and at a constant temperature this dependence is linear (Figure 94).

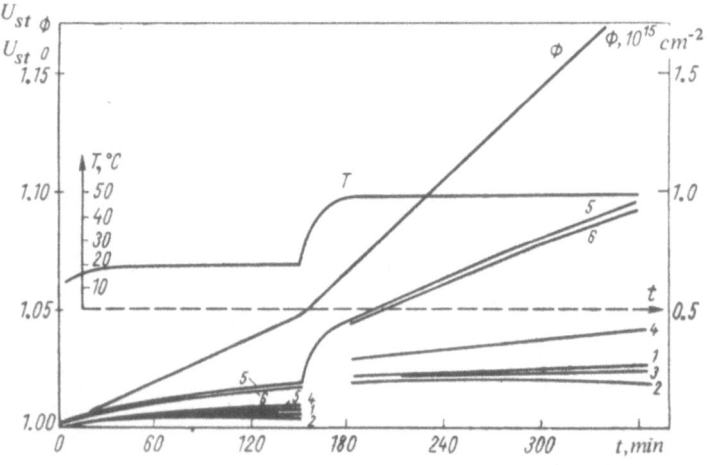

Figure 94. Changes in stabilization voltage (of low-voltage stabilitrons) upon irradiation with fast neutrons. In all cases $I_{rev} = 5$ mA. Curves 1 - 4: n-Si, $U_{st} = 8\text{-}10$ V; 5 and 6 - p-Si, $U_{st} = 10$ V.

Radiation changes not only the absolute value of the stabilization voltage but also the dynamic resistance in the breakdown region of the reverse characteristics. Since this resistance is determined mainly by the resistance of the diode base at high current densities ($>$1A/cm^2), the rate of its change should be proportional to the velocity of carriers. Such a proportionality has indeed been confirmed experimentally. The reverse V–I characteristics of stabilitrons after prolonged irradiation are shown in Figure 59. It should be pointed out that in the pre-breakdown region the characteristic becomes "soft" like in the case of high voltage diodes.

The forward characteristics indicate that stabilitrons, particularly low voltage ones, have a high radiation stability. The voltage drop monotonically decreases within a wide range of integrated fluxes as can be expected from diodes with either a thin base or made of very low resistivity materials (see equation (157) for the limiting current density $I_p^0 0$ at which $dU_g/d\Phi = 0$).

It has been found experimentally that in the range where U_s is a linear function of Φ neither the value nor the sign of the TCV undergo any considerable changes.

Let us now discuss briefly the radiation changes of the parameters of stabilitrons with a very low TCV, which consist of several diodes connected in series. The principle of such an arrangement is to compensate the positive TCV of the reverse breakdown voltage by a negative TCV of the forward voltage of several diodes. However, (see Figure 53) the forward voltage is changed by irradiation; according to /153/ – for this type of stabilitrons, it decreases monotonically at a rate of 10^{-15}-10^{-16}V (neutron/cm^2)$^{-1}$. Consequently, the stabilization voltage of such devices would continuously decrease during irradiation, at a rate about one order of magnitude higher than the voltage increase in ordinary stabilitrons. Therefore, such temperature-compensating stabilitrons should not be used under irradiation.

The above data on diodes and stabilitrons indicate that the radiation stabilities of common silicon devices, evaluated from the changes in the forward V–I characteristics may differ from each other by several orders of magnitude. Indeed, in p-i-n diodes with thick i-layers the forward voltage increases by several volts as a result of bombardment with a fast neutron fluence of only 10^{10}-10^{11}cm^{-2}. At the same time almost no changes in U_d have been found in lower-voltage stabilititrons irradiated with a fast neutron fluence $>10^{16}$cm^{-2}. Such a considerable difference between the resistivities of those devices is due, on one hand, to differences in the conductivity of the diode materials and on the other, to differences in the design characteristics or, more accurately, to differences in the base thickness.

It would seem that the radiation stability of diodes is not anymore a problem, since such diodes are stabilitrons whose radiation resistance is higher than that of, e.g., transistors. However, such diodes often cannot be used because of their low breakdown voltage, which is related to the conductivity of the material and the thickness of the base, i.e., with those factors that influence the changes in forward characteristics. The breakdown voltage can be increased by using materials of lower conductivity but this lowers the stability of the diode.

As we mentioned above, a radiation resistant diode (in the forward V–I characteristics) may in principle not have a high resistance. But since the thickness of the base begins to influence the breakdown voltage only when the base is so thin that the space charge region at high voltages overlaps with it completely, it is possible, by changing the base thickness and conductivity to choose such parameters and to design the diode in such a way that it will have a maximum possible resistance at a given breakdown voltage. This is, exactly, the purpose of proper design of radiation resistant diodes. Certain approaches to the solution of that problem are obvious from our earlier discussions.

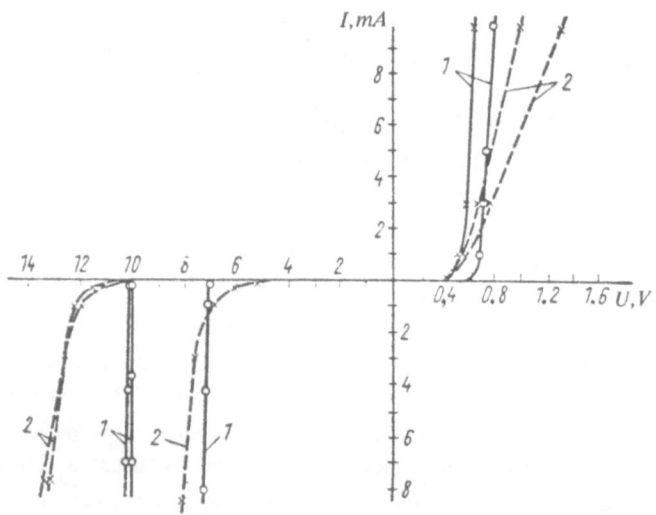

Figure 95. V–I characteristics of several low-voltage stabilitrons before (1) and after (2) irradiation with a fluence of $7 \cdot 10^{16}$ fast neutrons/cm^2

On the other hand data on the damage of diodes under irradiation, as reported above, will allow the designer to make the correct choice of diodes for prolonged service in the presence of radiation.

CHAPTER III

EFFECT OF PULSED RADIATION ON SEMICONDUCTOR DEVICES

1. Characteristics of Pulsed Irradiation of Semiconductor Devices.

Pulsed radiation refers to radiation which, over a small fraction of one second imparts to the irradiated target a high neutron fluence or a large γ-radiation dose; it is characterized by high radiation fluxes and by new effects that are not produced by prolonged continuous irradiation.

Pulsed radiation, like prolonged continuous radiation with fast neutrons, changes the characteristics of semiconductors and semiconductor devices. However, the final results of pulsed irradiation are more strongly influenced by the rate and flux density of radiation as well as by the amount of radiation (fluence or absorbed dose) than in the case of continuous irradiation. The processes generated by pulsed radiation in the bulk and on the surface of semiconductors do not stop with the interruption of irradiation but continue a certain time after it, leading to stable (at a given temperature) radiation defects which cause irreversible changes in the parameters of semiconductors and semiconductor devices. In the course of such non-equilibrium processes the defects may "ripen", e.g. Frenkel defects combine into more complex defects (vacancy-impurity complexes, divacancies, etc.) or may be annealed, filling the vacancies with interstitials or displaced atoms. The relationship between the rates of such conflicting processes is determined by the temperature of the semiconductor, by the concentration of impurities reacting with the defects and by other factors.

An investigation of the kinetics of formation and annealing of defects by pulsed irradiation of semiconductors and semiconductor devices is of great interest for the radiation physics of solids as it allows us to elucidate the production mechanism of defects stable at a given temperature.

In such investigations, pulsed radiation sources include linear electron accelerators /274/ which produce pulses lasting microseconds and fractions of a microsecond, and various pulsed reactors which produce pulses lasting from several tens of microseconds to several milliseconds /221,275/.

Rzewuski /276/, found that pulsed irradiation with accelerated electrons causes intense ionization and heating of semiconductors which interferes with the study of annealing processes that take place during the short period that follows the pulse and before the temperature and the steady-state concentration of carriers are recovered. The heating may be considerable (up to 20°C after a pulse of $5 \cdot 10^{13}$ electrons/cm^{-2}) even if the specimen is placed in liquid nitrogen. The rate of temperature recovery depends on the thermal conductivity of the material and on its thermal contact with the holder and with the heating unit.

The time of recovery after electron excitation is determined by the rate of recombination of excess carriers.

Neutron irradiation has in this case certain advantages and therefore it is widely used for the investigation of fast annealing of neutron-induced defects. A fluence of $5 \cdot 10^{12}$ neutron/cm^2, which can be produced by a 50 μsec pulse at halfwidth in the Godiva /221, 277/ or SPRF /278/ assemblies is sufficient to change by 5−10% the electrical resistivity of 10 ohm · cm germanium or silicon and also to change, by a factor of nearly 2, the base current transport coefficient of some transistors. The heating caused by such irradiation does not exceed several tenths of one degree.

The first experiments on transient processes of defect production and defect annealing in germanium were carried out by Stein /91/. Measurements of the conductivity and hole mobility showed that at temperatures near 200°K the resistivity of n-type germanium irradiated with neutron pulses increases both during and after the pulse. The recovery of the radiation-induced conductivity changes takes place mainly above 273°K.

Transistors have been used for the study of transient processes caused by pulsed radiation /272/, since they are less susceptible to ionization and may be operated at injection levels exceeding the excitation levels produced by residual γ-backgrounds.

Sander /280/, who studied annealing in transistors, compared the number of defects at a given moment with the number of defects remaining 10^3 sec after the pulse. The annealing coefficient was calculated from the ratio $(1 - \alpha(t))/(1 - \alpha(10^3))$ or $(1 - \alpha(t))/(1 - \alpha(\infty))$; it is evident that its value was always greater than unity.

Sander and Gregory /281/ found that for n-p-n planar epitaxial 2N914 silicon transistors irradiated with neutron pulses ($\Phi = 1.8 \cdot 10^{13}$

neutron/cm^2, $E_n > 10$ keV) the annealing coefficient corresponding to a 10^{-4} sec pulse decreased with increasing temperature. The ratio of the 10^{-4} to 10^3 sec. coefficients increased from 2 to 5 when the temperature was reduced from 248 to $213°K$. The annealing processes had two stages: a low-temperature stage and a second stage which started at the moment when the annealing coefficient reached a value of 2.5.

Binder *et al.* /282/ found that the first stage of annealing can be represented by a kinetic equation of the first order, which means that the concentration of defects decreases exponentially with time. Therefore, they assumed that this stage is associated with recombination of close-lying neighboring defects. On the other hand, the time dependence of the second stage corresponds to a diffusion-limited annealing process /282/.

Intense ionization of materials and the production of large transient currents and emfs in semiconductor devices is another important effect of pulse irradiation. The ionization is caused mainly by absorption of γ-radiation. The contribution of neutrons to ionization during combined gamma and neutron irradiation in pulsed reactors does not exceed 0.2% for germanium and 7% for silicon /283/.

The absorption of X-ray or γ-rays as a result of the photo-effect, Compton effect or electron-positron pair formation produces free electrons with an energy sufficient to initiate cascade multiplication processes as a result of impact ionization. The absorption energy is used for the ionization, i.e. for the formation of free holes and electrons in the valence band and in the conduction band and also for the heating of the semiconductors. The mean energy ε consumed in the production of a single electron-hole pair during the absorption of high-energy quanta, is much greater than the forbidden gap width. The experimental values of ε (3.6 and 3.0 eV for silicon and germanium respectively) /13/ agree well with values calculated on the basis of the Shockley theory /181/. By using those values of the mean energy of pair production, the density of germanium and silicon and the definition of 1 rad (as the amount of ionizing radiation energy absorbed by a unit of mass i.e. 1 rad = 100 erg/g) we can show that X-rays and γ-radiation at a dose rate of 1 rad/sec produces $4.3 \cdot 10^{13}$ pairs/(cm^3·sec) of carriers in silicon and about $1.1 \cdot 10^{14}$ pairs/(cm^3·sec) of carriers in germanium.

If diodes and transistors are irradiated with γ-rays, part of the carriers produced in the vicinity of the *p-n* junction will be separated by the junction and as a result will produce either a voltage or a current, if the device is connected to a closed circuit. At high ionizing radiation current densities the radiation-induced transient currents can reach the value of the working currents of the devices or even exceed them, thus upsetting the operation of electronic devices.

2. Ionization current in p-n junctions.

In order to describe the transient processes generated by radiation pulses in a *p-n* junction, let us consider the concentration distribution of non-equilibrium carriers produced by irradiation. The concentration distribution of non-equlibrium carriers before irradiation in reverse biased *p-n* junctions is shown in Figure 96a (continuous line). In this

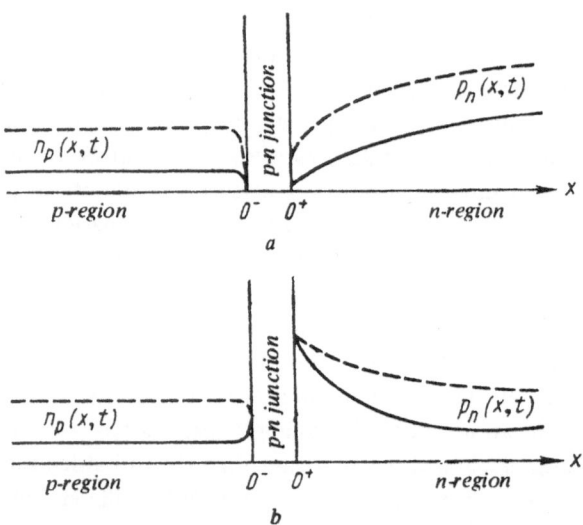

Figure 96. Concentration distribution of minority carriers in the vicinity of *p-n* junctions before (continuous curve) and during irradiation (dash curve): *a*—reverse biased junction; *b*—forward biased junction.

case irradiation produces carrier pairs uniformly distributed within the bulk. The carriers produced in the space charge region will be accelerated by the junction field and collected by the *p* and *n* regions in the course

232

of nanoseconds, i.e. almost immediately after the start of irradiation. Therefore, the current produced by the motion of these carriers does not lag behind the radiation pulses and is called the instantaneous component of ionization current $i_m(t)$.

Carriers produced outside the p-n junctions raise the concentration of minority carriers. This is shown by the dash line, in Figure 96, which represents the concentration distribution of carriers at the end of a very short radiation pulse. Since the concentration gradient of minority carriers in the vicinity of the junction is increased by irradiation, the current passing through the junction is gradually increased and maintained afterwards at a high value by the carriers produced in the bulk limited by the diffusion length. Carriers produced at a distance greater than that limit cannot reach the junctions because of recombination and therefore they do not contribute to the ionization current. Since the carriers produced in the bulk limited by the diffusion length need a certain time to reach the junction, the current produced by them lags behind the radiation impulse. This current is called the delayed component of the ionization current $I_d(t)$.

Those two components increase the initial reverse current which flows through the junction from the n to the p field. The duration of this increase depends on the duration of the radiation pulse and of the time of recovery of the equilibrium state.

Similar transient processes occur in semiconductor devices with forward biased p-n junctions. However, since the forward bias does not entirely compensate the intrinsic field of the p-n junction, the instantaneous component of the ionization current will flow as previously from the n to the p-regions, i.e. in a direction opposite to the forward current. Figure 96b shows that carriers produced in the n- and p- regions outside the junction reduce both the concentration gradient of minority carriers injected by the junction and the forward current. Thus, the current generated at a given moment of irradiation can be determined by adding the normal forward current to the two components of the ionization current.

Let us now determine the value and the time dependence of the two components of the ionization current in a device with a single p-n junction, assuming that:

1. The device has a one-dimensional structure.
2. The length of the p- and n-regions. is much greater than the diffusion length of the corresponding minority carriers.
3. The radiation dose rate is smaller than that required to modulate

the conductivity of the p- and n-regions
4. The p- and n-regions of the device are uniformly doped and there is therefore a weak field or no field at all in the bulk, except the p-n junction.
5. The voltage across the p-n junction is constant during the transient processes and the device with a reverse-biased p-n junction does not become saturated.

If those conditions are satisfied, the instantaneous component of the ionization current will be determined by the width of the barrier layer w_{pn} and by the rate of pair production $g(t)$:

$$i_m(t) = q A_{pn} w_{pn} g(t),\qquad (194)$$

while the delayed component will be determined by the concentration gradient of minority carriers in the p and n regions:

$$i_d(t) = q A_{pn} \left[D_p \frac{\partial p_n(x,\,t)}{\partial x}\bigg|_{x=+0} - D_n \frac{\partial n_p(x,\,t)}{\partial x}\bigg|_{x=-0} \right]. \qquad (195)$$

The corresponding concentration gradients of carriers can be determined by solving the differential continuity equation for p and n regions for the following starting and terminal conditions:

$$p_n(+0,\,t) = p_{n0} \exp(qU/kT), \qquad n_p(-0,\,t) = n_{p0} \exp(qU/kT),$$
$$\lim_{x \to +\infty} |p_n(x,\,t)| < \infty, \qquad \lim_{x \to -\infty} |n_p(x,\,t)| < \infty.$$

where $p_n(x,0)$ and $n_p(x,0)$ correspond to the equilibrium distributions of the reverse current U_{rev} applied to the junction.

Let us solve the differential continuity equation for the n-region:

$$\frac{\partial p_n(x,\,t)}{\partial t} = D_p \frac{\partial^2 p_n(x,\,t)}{\partial x^2} - \frac{p_n(x,\,t) - p_{n0}}{\tau_p} + g(t). \qquad (196)$$

At a constant U_{rev}, the above linear equation with constant coefficients can be conveniently solved by an operator transformation. A similar solution can be obtained for the p-region. Without describing the calculations we can write the results in an operator form.*

---

* About the solution of partial derivatives by the operational method, see /284/.

$$P_n(x, s) = \frac{G(s)}{s + 1/\tau_p}\left[1 - \exp\left(-x\sqrt{(s + 1/\tau_p)/D_p}\right)\right] + p_n(x, 0),$$

$$(197)$$

where $G(s) = g(t)$.

By substituting into (195) the expression for $P_n(x, s)$ and the similar expression for $N_p(x, s)$ we obtain an operational expression for the ionization current:

$$I_i(s) = I_d(s) + I_m(s) =$$
$$= qA_{pn}G(s)\left[w_{pn} + \frac{\sqrt{D_p\tau_p}}{\sqrt{1 + s\tau_p}} + \frac{\sqrt{D_n\tau_n}}{\sqrt{1 + s\tau_n}}\right] \qquad (198)$$

For particular cases the inverse transformation of the function $g(t)$ yields:

(a)-a stepwise function with an amplitude G at the moment $t=0$:

$$i_i(t) = qA_{pn}G\left[w_{pn} + \sqrt{D_p\tau_p}\,\mathrm{erf}\left(\sqrt{t/\tau_p}\right) + \sqrt{D_n\tau_n}\,\mathrm{erf}\left(\sqrt{t/\tau_n}\right)\right];$$
$$(199)$$

(b)-the same at $t \to \infty$ (quasiequilibrium current)

$$I_i = qA_{pn}G\left[w_{pn} + L_p + L_n\right], \qquad (200)$$

where $\qquad L_p = \sqrt{D_p\tau_p}\ $ and $\ L_n = \sqrt{D_n\tau_n};$

(c)-for a rectangular pulse of amplitude G and duration t_i we have:

$$i_i(t) = qA_{pn}G\left[w_{pn} + \sqrt{D_p\tau_p}\,\mathrm{erf}\left(\sqrt{t/\tau_p}\right) + \sqrt{D_n\tau_n}\,\mathrm{erf}\left(\sqrt{t/\tau_n}\right)\right]$$
$$\text{for} \quad 0 \leqslant t \leqslant t_i; \qquad (201a)$$
$$i_i(t) = qA_{pn}G\left\{\sqrt{D_p\tau_p}\left[\mathrm{erf}\left(\sqrt{t/\tau_p}\right) - \mathrm{erf}\left(\sqrt{(t - t_i)/\tau_p}\right)\right] +\right.$$
$$\left. + \sqrt{D_n\tau_n}\left[\mathrm{erf}\left(\sqrt{t/\tau_n}\right) - \mathrm{erf}\left(\sqrt{(t - t_i)/\tau_p}\right)\right]\right\}$$
$$\text{for} \quad t > t_i, \qquad (201b)$$

(d)-the extreme case of the previous problem at $t_i \to 0$

$$i_i(t) = qA_{pn}Gw_{pn} \qquad 0 \leqslant t \leqslant t_i; \qquad (202a)$$
$$i_i(t) = qA_{pn}Gt_i\left[\frac{\sqrt{D_p}\,e^{-t/\tau_p} + \sqrt{D_n}\,e^{-t/\tau_n}}{\sqrt{\pi t}}\right] \qquad (202b)$$
$$\text{for} \quad t > t_i;$$

(e)-A general case, i.e., arbitrary function $g(t)$

$$i_n(t) = qA_{pn}\left\{w_{pn}g(t) + \right.\tag{203}$$

$$\left. + \int_0^t g(t-\lambda)\left[\frac{\sqrt{D_p}\,e^{-\lambda/\tau_p} + \sqrt{D_n}\,e^{-\lambda/\tau_n}}{\sqrt{\pi\lambda}}\right]d\lambda\right\}.$$

The assumption about a weak electric field in the bulk of the material outside the p-n junction is not valid for devices with a forward biased junction. However, if we assume that the ionization current and the injection levels are smaller than the forward current and the concentration of majority carriers, then the existing fields would not change much and a similar analysis in this case would yield (for a steady state /285/):

$$I_i = qA_{pn}G\left[\tau_p\left(\sqrt{\frac{D_p}{\tau_p} + \frac{\mu_p^2E_p^2}{4}} - \frac{\mu_pE_p}{2}\right) + \right.$$

$$\left. + \tau_n\left(\sqrt{\frac{D_n}{\tau_n} + \frac{\mu_n^2E_n^2}{4}} - \frac{\mu_nE_n}{2}\right) + w_{pn}\right],\tag{204}$$

where E_p and E_n are the electric fields in the p and n regions respectively; equation (204) may now be written as follows:

$$I_i = qA_{pn}w_{pn}G\tag{205}$$

for strond fields corresponding to forward bias.

3. Equivalent Diode Circuit Modelling Diode Response to Ionizing Radiation Pulses.

We discussed above the ionization currents in devices with a single p-n junction, which are idealized diodes. In the case of real diodes used in radio and electronic circuits similar simplifications would not be always valid. Although this may not greatly affect the final results, the use of $i_i(t)$ for idealized diodes may result in substantial errors in the evaluation of the magnitude and the nature of real transient processes.

Idealized diodes correspond most closely to diodes with a graded p-n junction made by crystal pulling from a melt or by dopant diffusion. In alloyed diodes, (unlike diodes obtained by crystal growth or diffused diodes) one of the regions is strongly doped and contains minority carriers with short lifetimes. Therefore the corresponding term

236

in equations for $i_i(t)$ may be disregarded. Such diodes have, usually, a strongly developed peripheral region of the base crystal and therefore their effective volume, which determines the equilibrium value of the delayed component of the ionization current i_d, is larger than the product of the junction surface area A_{pn} and the diffusion length of minority carriers. Because of differences between the rates of recombination of minority carriers in the bulk and on the surface, the mean carrier collection times from the bulk of the crystal differ from the carrier collection time from the peripheral surface regions. As a result, the front of the delayed component pulse of ionization current could not be represented by the function $erf\sqrt{t/\tau}$.

The conductivity of the base material of diodes is often modulated and therefore the p and n regions of such diodes are not much longer than the diffusion length. The lifetime τ in equations for $i_i(t)$ is in such a case replaced by the diffusion time of carriers from the ohmic contact to the p-n junction: however, the equation would depend on the conditions of recombination at the contact.

In pulsed diodes, where the lifetime of the minority carriers can be reduced to a very small value, the delayed component $i_d(t)$ which usually makes the greatest contribution to $i_i(t)$ may be smaller than the instantaneous component $i_n(t)$. Under such conditions, the ionization current pulses would to a considerable degree repeat the radiation pulses and their amplitude would depend not only on the radiation dose rate but also on the reverse voltage applied to the diode.

In real circuits with biased diodes the reverse voltage across the junction does not remain constant during pulsed radiation and therefore it changes the width of the barrier layer of the junction w_{pn} and its charge capacitance C_{pn}. Therefore, both the ionization current and the voltage of diodes are determined not only by the parameters of the radiation pulses but also by the external circuit. In an extreme case when the radiation doses are high and the current of the reverse biased diode circuit is limited, the polarity of the p-n junction is reversed and the junction becomes forward biased. The diodes will remain in such a state of saturation for a certain time after irradiation is stopped.

The response of many real diode circuits to pulsed ionization can be modeled by an equivalent circuit (Figure 97) based on concepts developed for transistors /286/.

According to those concepts the diode current is determined by the charge Q, accumulated in the bulk of the p and n regions. The storage

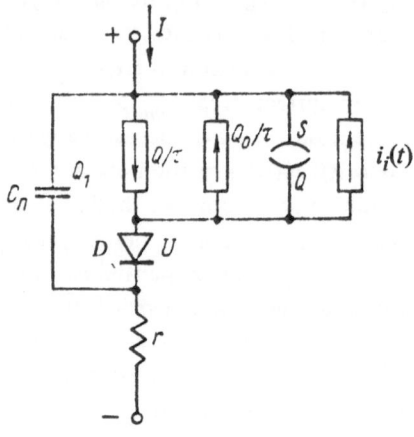

Figure 97. Equivalent circuit of a diode, modelling its response to ionizing radiation pulses.

element S in the equivalent circuit reflects the capacities of p and n regions to accumulate identical amounts of electrons and holes and to remain neutral* i.e. to produce no electric field. This storage element can be regarded as an infinite capacitance. The charge Q is an exponential function of the voltage U across an ideal diode $(p$-n junction of a real diode):

$$Q = Q_0 \exp (qU/kT), \qquad (206)$$

where Q_0 is the charge at $U=0$.

The total current flowing through a diode is determined by three sources of current, Q/τ, Q_0/τ and $i_i(t)$. The current produced by the first source is determined by the voltage applied to the junction. The current from the second source corresponds to the saturation current and the current from the third source is equal to the ionization current in an idealized diode $i_i(t)$. The capacitor C_{pn}, the charge Q and the resistor r reflect the charge capacitance of the barrier layers of p-n junctions in real diodes, the charge in and the resistance of the crystal respectively. Neither C_{pn} nor r remain constant if the voltage and the

---

* At a low injection level and a relatively small concentration of minority carriers, i.e. $\Delta n = \Delta p \ll n_n$ or $\Delta n = \Delta n = \ll p_p$.

238

current through the diode are changed; however, the use of mean values in the calculations yields satisfactory results.

The differential equation of charge preservation for the above equilibrium circuit is as follows:

$$\frac{dQ}{dt} + \frac{dQ_1}{dt} + \frac{Q}{\tau} - \frac{Q_0}{\tau} - I - i_i(t) = 0. \tag{207}$$

If the forward and reverse voltages on the diode are constant, we have:

(a)-at U_{rev}=const:

$$Q = 0, \qquad \frac{dQ}{dt} = \frac{dQ_1}{dt} = 0,$$

where U_{rev} is the reverse voltage. The current in the external circuit would be:

$$I = -\left[\frac{Q_0}{\tau} + i_i(t)\right] = -[I_s + i_i(t)], \tag{208}$$

since by definition $Q_0/\tau = I_s$ (I_s is the saturation current of the diode):

(b)-at U_{for} = const we have:

$$Q = Q_0 \exp(qU/kT), \qquad \frac{dQ}{dt} = \frac{dQ_1}{dt} = 0;$$

$$I = \frac{Q}{\tau} - \frac{Q_0}{\tau} - i_i(t) = I_s[\exp(qU_{pr}/kT) - 1] - i_i(t). \tag{209}$$

Often the reverse current is limited by the external resistance R to a value $I_{rev\ max} = E/R$; therefore the diode is saturated at high dose rates. For reverse biased p-n junctions we may assume that $Q = 0$ and $dQ/dt \ll dQ_1/dt$. The error introduced by making this assumption is considerable only at reverse voltages which differ from zero by several kT/q. The equation for charge preservation in the first stage of the transient process ($U_{rev} < 0$) at C_{pn}=const should be written:

$$\frac{dQ_1}{dt} - \frac{Q_0}{\tau} - I - i_i(t) = 0, \tag{210}$$

but since

$$\frac{dQ_1}{dt} = C_{pn}\frac{dU}{dt}, \; Q_0/\tau = I_s \text{ and } I = -\frac{U+E}{R}, \text{ we have}$$

$$\frac{dU}{dt} + \frac{U}{RC_{pn}} = -\frac{E}{RC_{pn}} + \frac{1}{C_{pn}}[I_s + i_i(t)] \qquad (211)$$

the initial condition being that $U(0) = -E + I_s R$.

Solution of this equation by the operator method gives:

$$U(s) = -\frac{E - I_s R}{s} + \frac{I_i(s)}{C_{pn}(s + 1/RC_{pn})}, \qquad (212)$$

where s is the argument of function transform. For a rectangular radiation pulse of amplitude G and duration $t_i (t_i > \tau$ and $t_i > RC_{pn})$ the front of the ionizing current pulse will be $i_i(t) = I_i erf\sqrt{t/\tau}$ where $I_i = QA_{pn}LG$ (here, as before, A_{pn} denotes the junction surface area).

After transformation we obtain

$$I_i(s) = \frac{I_i}{\sqrt{\tau s}\,\sqrt{s + 1/\tau}}. \qquad (213)$$

By substituting equation (213) in (212) and performing inverse transformation we obtain the following equation for the diode voltage:

$$U(t) = -(E - I_s R) + I_i R\left[erf\sqrt{t/\tau} - \qquad (214) \right.$$

$$\left. - \frac{e^{-t/RC_{pn}} erf\sqrt{\frac{t}{\tau}\left(1 - \frac{\tau}{hC_{pn}}\right)}}{\sqrt{1 - t/RC_{pn}}} \right].$$

Figure 98 shows graphically the function:

$$F(t) = erf\sqrt{t/\tau} - \frac{e^{-t/RC_{pn}} erf\sqrt{\frac{t}{\tau}\left(1 - \frac{\tau}{RC_{pn}}\right)}}{\sqrt{1 - t/RC_{pn}}}$$

for various τ/RC_{pn}.

A saturation current is established at the moment t_1, which is determined from the equation: $F_{(t_1)} = \frac{E - I_s R}{I_i R}$

During the second stage of the transition process the forward bias voltage increases from zero. The solution of equation (207) for this period under the condition that

240

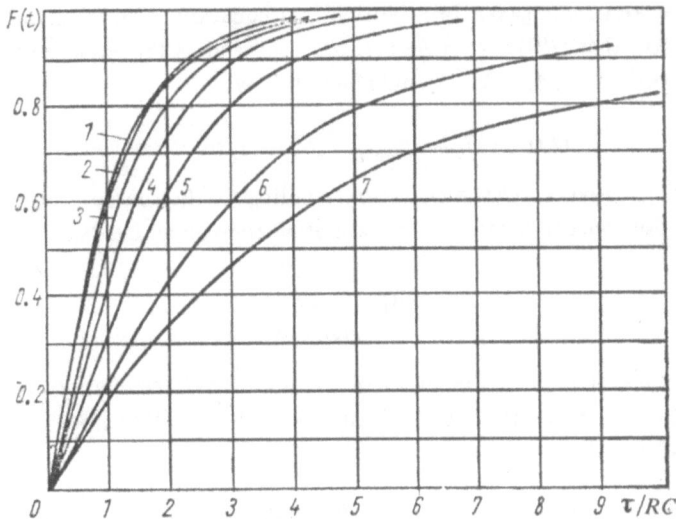

Figure 98. $F(t)$ plot for various τ/RC_{pn} values:
1 - 0.1; 2 - 0.2; 3 - 0.5; 4 - 1.0; 5 - 2.0; 6 - 5.0; 7 - 100.

$$\frac{dQ_1}{dt} < \frac{dQ}{dt}, \quad Q(0) = 0 \text{ and } i_i(t) = I_i \operatorname{erf} \sqrt{(t+t_1)/\tau},$$

would be as follows:

$$Q(t) = Q_0 - \tau I_{rev,\,max} \ (1 - e^{-t/\tau}) + I_i \tau \times$$
$$\times \left[\operatorname{erf} \sqrt{\frac{t+t_1}{\tau}} - \operatorname{erf} \sqrt{\frac{t_1}{\tau}}\, e^{-t/\tau} \right] -$$
$$-I_i \frac{2\sqrt{\tau}}{\sqrt{\pi}}\, e^{-\frac{(t+t_1)}{\tau}} [\sqrt{t_1+t} - \sqrt{t_1}]. \tag{215}$$

The equilibrium value of $Q(t)$ at $t \to \infty$ is:

$$Q(t)_{t\to\infty} = Q_0 + I_i \tau - I_{rev,\,max}\, \tau \tag{216}$$

That value corresponds to an equilibrium forward voltage on the diode:

$$U_{for} = \frac{kT}{q} \ln 1 + \frac{I_i - I_{rev,\,max}}{I_s} \tag{217}$$

241

For the third stage of the transient process (or dissipation of accumulated charges) the solution of equation (207) under the condition that $Q(0) = Q_0 + I_i\tau + I_{rev.\,max}\tau$ and in the absence of any ionization current $[I_i(t)=0]$ would be as follows:

$$Q(t) = Q_0 - I_{rev,\,max} \cdot \tau + I_i \tau e^{-t/\tau}. \qquad (218)$$

Using equation $Q(t) = Q_0$, corresponding to the diode output at saturation, we can determine the saturation time under irradiation

$$t_{sat} = \tau \ln \frac{I_i}{I_{rev,\,max.}}. \qquad (219)$$

The last stage in the recovery of the reverse voltage does not differ, in principle, from that for ordinary switching of the diode from a reverse to a forward bias /287, 288/.

4. Response of Transistors to Ionizing Radiation Pulses and Equivalent Circuits.

The development of transient processes in transistors under the influence of pulsed radiation is discussed below using as an example a device with a *p-n-p* structure, in a common emitter circuit operating in an active mode, i.e., with a forward biased emitter junction and a reverse biased collector junction. During the initial stage, the transient process in such transistors is similar to that in diodes. Electrons and holes generated by the radiation diffuse into the *p-n* junctions and are accumulated in the *p-* and *n*-regions respectively. Their motion through the junctions produces an ordinary ionization current, with instantaneous and delayed components.

During the first stage of this process the uncompensated charge of majority carriers (electrons) is accumulated in the base of the transistor. This negative charge reduces the height of the potential barrier of the emitter junction, causing additional injection of holes from the emitter into the base. The injected holes neutralize the non-equilibrium charge of the electrons and, by diffusing to the collector, generate a large transistor current (which flows during the time a non-equilibrium charge exists in the base).

In alloyed-diffused and in diffused transistors the diffusion length of minority carriers exceeds the base region width and, therefore, a charge is accumulated mainly because of the flow of electrons from the collector. On the other hand, because of the very short diffusion length

242

of minority carriers in highly-doped emitters and collectors of alloyed transistors, electrons are generated mainly in the base and its peripheral regions.

An accurate description of the magnitude and nature of transient processes in transistors must take into account factors such as the redistribution of charges, the rate of recombination processes in the bulk and on the surface, the real geometry, and many other factors which are hard to predict for encapsulated devices. Some of those factors are of minor importance. In order to determine the more important factors, equation for non-equilibrium collector currents are based on certain assumptions which allow us not only to determine the mechanism of the process but also to make approximate calculations that are confirmed experimentally.

The basic assumption is that the equivalent transistor curcuit can be used as a charge-controlled device /286/. It is also assumed that:

(*a*) The transistor has unidimensional geometry.

(*b*) The diffusion or drift of carriers through the base is brief enough to disregard changes in dose rate and in the base and collector currents during that time.

(*c*) There is either no electric field in the base or there is a field corresponding to the function $E(x) = D\dfrac{dF(x)}{dx}\Big/ \mu F(x)$ where μ is the carrier mobility and $F(x)$ is an arbitrary function.

(*d*) The base current and the collector voltage are constant, i.e., the influence of the capacitance of the collector junction and of the external circuit on the nature of the process are disregarded.

(*e*) The capacitances of barrier layers in the emitter and collector junctions are smaller than the diffusion capacitance of the base. This condition allows us to represent transient processes by a linear differential equation with constant coefficients and to solve it with the aid of tabulated functions (in some cases it is impossible to disregard the capacitance of the barrier layer of the emitter; this is discussed below).

During irradiation the charge of majority carriers in the base is influenced by the ionization current of the collector junction, by the emitter junction ionization current, by the current generated by pair production in the base, and by the normal base current. Since the emitter junction ionization is usually weak and the current generated by pair production in the base may be regarded as an instantaneous component of the collector junction ionization current, the equivalent

243

circuit of a charge controlled transistor /286/ (Figure 99) contains only one source of ionization current $i_i(t)$. Therefore, for an arbitrary $g(t)$ that current could be descried as follows:

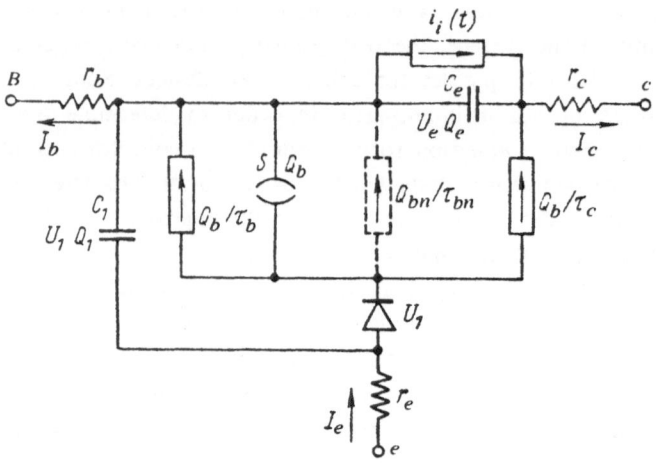

Figure 99. Equivalent circuit of charge-controlled p-n-p transistor.

$$i_i(t) = qA_c \left\{ (w_b + w_{pnc}) g(t) + \int_0^t g(t - \lambda) \left[\frac{\sqrt{D_{ce}}}{\sqrt{\pi \lambda}} e^{-\lambda/\tau_c} \right] dx \right\}. \tag{220}$$

where w_b is the base thickness. The subscript "c" shows that the parameter is related to the collector or to the collector junction.

In the past, it was assumed that because of the short transit time of carriers through the base all minority carriers produced by radiation are swiftly gathered by the collector and therefore, the corresponding current is described in equation (220) as an instantaneous component. In reality, the current generated by pair production in the base has, (particularly in alloy transistors) also a delayed component because of collection of carriers from the peripheral region of the base.

Using Kirchhoff's law the following equation is written for point B

$$-I_b - i_i(t) + \frac{Q_b}{\tau_b} + \frac{dQ_b}{dt} + \frac{dQ_1}{dt} - \frac{dQ_3}{dt} = 0. \tag{221}$$

Disregarding the influence of the junction capacitances C_{eb} and C_{cb} we obtain:

$$\frac{dQ_b}{dt} + \frac{Q_b}{\tau_b} - I_b - i_i(t) = 0. \quad \cdot \tag{222}$$

Similarly for point K we obtain:

$$i_i(t) = i_i(t) + \frac{Q_b(t)}{\tau_c}, \tag{223}$$

or since $\tau_c = \tau_b/B$ /289/, (where B is the base current transport coefficient or gain factor) we obtain:

$$i_c(t) = i_i(t) + \frac{BQ_b(t)}{\tau_b}. \tag{224}$$

Solution of equations (222) and (224) with the aid of operators gives:

$$I_i(s) = I_i(s) + \frac{I_b B}{\tau_b s(s+1/\tau_b)} + \frac{BI_i(s)}{\tau_b(s+1/\tau_b)} + \frac{Q_b(0)B}{\tau_b(s+1/\tau_b)}. \tag{225}$$

In some cases inverse transformation of $g(t)$ yields:
(a) for a stepwise function of G at a moment $t = 0$:

$$+\sqrt{D_c\tau_c}\,\mathrm{erf}\,\sqrt{t/\tau_c} -$$
$$-\sqrt{\frac{D_c}{1/\tau_c - 1/\tau_b}}\,e^{-t/\tau_b}\,\mathrm{erf}\,\sqrt{t(1/\tau_c - 1/\tau_b)}\Big]\Big\}; \tag{226}$$

$$i_c(t) = qA_cG\Big\{w_b + w_{pnc} + \sqrt{D_c\tau_c}\,\mathrm{erf}\,\sqrt{t/\tau_c} + $$
$$+ B\Big[(w_b + w_{pnc})(1 - e^{-t/\tau_b}) + $$

(b) for the same function at $t \to \infty$ (quasi-equilibrium current):

$$I_i = qA_cG[w_b + w_{pnc} + \sqrt{D_c\tau_c} + B(w_b + w_{pnc} + \sqrt{D_c\tau_c})]; \tag{227}$$

(c) if $g(t)$ represents a rectangular pulse of amplitude G and duration t_i the equation can be solved as for ('a'), by superposing two step-wise step functions, of opposite signs and with a delay time t_i.

(d) $g(t)$ is an arbitrary function which acts over a time t_i smaller than all time constants of the transistor:

245

$$i_c(t) = qA_c(w_b + w_{pnc})g(t) +$$

$$+ qA_c \frac{B}{\tau_b}(w_b + w_{pnc}) \int_0^t g(\lambda)\, d(\lambda) \qquad (228)$$

for a time range $0 \leqslant t \leqslant t_i$,

$$i_c(t) = qA_c \int_0^{t_i} g(\lambda)\, d(\lambda) \left\{ \frac{\sqrt{D_c}}{\sqrt{\pi t}} e^{-t/\tau_c} + \qquad (229) \right.$$

$$\left. + \frac{B}{\tau_b} e^{-t/\tau_b} \left[w_b + w_{pnc} + \frac{\sqrt{D_c}}{\sqrt{1/\tau_c - 1}\, \tau_b} \operatorname{erf} \sqrt{t(1/\tau_c - 1/\tau_b)} \right] \right\}$$

for $t > t_i$.

According to /228/, the rise time of the ionization current in alloyed transistors (except the peripheral region) is equal to the duration of the radiation pulse, and the decay of the current pulse is an exponential function with a time constant equal to the lifetime of minority carriers in the base τ_b.

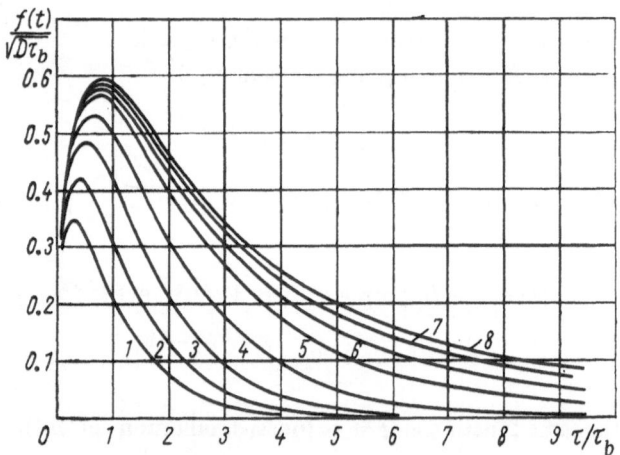

Figure 100. Plots of the normalized function $f(t)$ for various values of τ/τ_b; 1 - 1/4; 2 - 1/2; 3 - 1; 4 - 2; 5 -4; 6 -6; 7 - 8; 8 -10.

In diffused and alloyed-diffused transistors the duration of the current pulse depends on the ratio of the lifetimes of minority carriers in the base to that in the collector. Figure 100 shows graphically the normalized function:

$$f(t) = \sqrt{\frac{D}{1/\tau - 1/\tau_b}}\, e^{-t/\tau} b\, \text{erf} \sqrt{t\,(1/\tau - 1/\tau_b)}$$

for various values of τ/τ_b which may be used to determine the collector current pulses in such transistors.

Transistors operated in the vicinity of the cutoff region, i.e. at microamperes, or high frequency transistors with a very narrow base may have a diffusion capacitance smaller than the capacitance of the emitter junction. The excess charge of majority carriers generated in the base by radiation is compensated mainly by ionized atoms in the barrier layer of the emitter rather than by minority carriers injected from the emitter; therefore most of the above conclusions would not be valid in this case /290/.

The voltage change on the emitter junction of transistors is an approximately linear function of the charge

$$\Delta Q_1 = C_1 \Delta U_1 + \Delta C_1 U_1 \tag{230}$$

provided that the conditions of transistor operation are such that C_1 remains almost constant ($\Delta C_1 \approx 0$).

Since the current flowing through the junction is determined by the function $I = I_0 [exp(qU_1/mkT)^{-1}]$ (in the limiting case of $Q_1 \gg Q_b$) the maximum value of the collector current is an exponential function of the radiation dose (at small t_i). The changes in C_1 as a result of a change in the voltage U_1 have little effect on the final results.

The nonlinear differential equations representing the above non-equilibrium processes have no general solution, but must be solved individually with the aid of analog or digital computers /289/.

Earlier solutions of differential equations were based on the assumption that there is no electric field in the collector region and, therefore, the delayed component of the ionizing collector current is generated solely by the diffusion of carriers. At high radiation dose rates ($\varphi_\gamma > 10^7 - 10^8$ rad/sec) this assumption may not be valid for transistors with high-resistivity collector regions and the current may produce an accelerating electric field that will influence the nature of the process. Indeed, all other conditions being equal, the electric field, which causes drift of minority carriers from the collector to the base, and the diffusion of carriers, increase the charge of the base. This increases the collector current and the electric field strength. This mechanism of current multiplication in the collector, which is similar to positive feedback coupling, may lead, particularly in transistors with a high gain factor, to instability characterized by an exponential increase in the collector current with time. Stability is reestablished if recombinations decrease the concentration of minority carriers in the collector region to a level at which the carrier current injected into the base cannot be sustained by the electric field.

A similar instability is illustrated by an oscillogram of the collector current of a diffused *n-p-n* silicon 2N1051 mesa transistor (Figure 101).

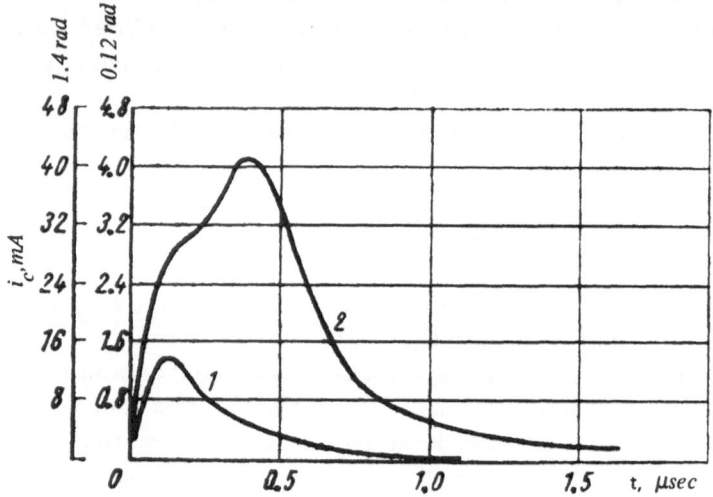

Figure 101. Collector current pulse in a 2N1051 transistor for an absorbed dose of 0.12 (1) and 1.4 (2) rad (duration of of radiation pulse 0.1 μsec).

Small absorbed radiation doses (0.12 rad) cause a transient process which is quite common for transistors of this type but if the doses are increased to 1.4 rad the field produced by the large collector current is not sufficient to initiate a multiplication process. Since the electric field is proportional to the collector current which is a function of this field, the mathematical model which describes the transient process with multiplication is not a linear one and can be solved only by numerical methods. The exact computation of the process is difficult since the collector current depends on the operating conditions of the transistor and is not necessarily related (by a linear relationship) to the dose or the radiation dose rate /285/.

If the collector current is limited by the external resistance or by the resistance of the collector region, the transistor can be saturated by strong irradiation. The saturation of alloy transistors involves accumulation of an excess charge of minority carriers in the base region. In diffused transistors and in alloyed-diffused transistors, saturation leads to accumulation of the excess charge mainly in the collector region which has relatively high resistivity and thickness and which ordinarily contains minority carriers with a long lifetime.

A charge-controlled equivalent transistor circuit can be used even when the transistor is saturated /289/. In that case the circuit must contain an additional current generator $I = Q_{bn}/\tau_{bn}$ (dash line in Figure 99) and the charge accumulated in the base is divided into two parts: Q_{ba}-normal base charge controlled by the transistor in the active region and Q_{bn}-base charge associated with saturation.

Thus, depending on whether the transistor is in an active or in a saturated state, we would have:

$$\left.\begin{array}{l} Q_b = Q_{ba} \ \ \textit{in the active region} \\ Q_b = Q_{ba\ max.} \ \ \ Q_{bn} = 0 \ \ \textit{at the saturation limit} \\ Q_b = Q_{ba\ max.} + Q_{bn} \ \ \ \textit{at saturation} \end{array}\right\} \tag{231}$$

Both charges are closely associated with the voltages U_i and U_d across the junction.

$$Q_{ba\ max.} + Q_{bn}/2 = Q_{be} \ \exp(qU_1/kT), \tag{232}$$

$$Q_{bn}/2 = Q_{bc} \ \exp(-qU_d/kT), \tag{233}$$

where Q_{be} is equal to the charge of minority carriers in the base, when their concentration at the emitter corresponds to the thermal equilibrium value and their concentration at the collector is equal to zero. Q_{bc} is related to the reverse collector current I_{co}. If the surface areas of the emitter and the collector are identical we obtain $Q_{be} = Q_{bc}$.

As an example let us determine (using the equivalent circuit in Figure 99) the time of dissipation of the charge accumulated in alloyed-transistors after the termination of the radiation pulse. The equation for charge preservation during saturation can be written as follows:

$$\frac{dQ_b}{dt} = \frac{dQ_{bn}}{dt} = -\frac{Q_{bn}}{\tau_n} - \frac{Q_{ba\,max.}}{\tau_b} + I_b + i_i\,(t) \qquad (234)$$

or

$$\frac{dQ_{bn}}{dt} + \frac{dQ_{bn}}{\tau_n} + \frac{I_{c(sat)}}{B} - I_b - I_i\,\frac{1+B}{B} = 0,$$

since

$$\frac{Q_{ba\,max}}{\tau_b} = \frac{I_{c(sat)} - I_i}{B},$$

where $I_{c(sat)}$ is the collector current in a state of saturation and $I_i -$ is the steady-state value of the ionization current flowing through the collector junction. Since in such transistors the ionizing current consists of the instantaneous component only*, immediately after the irradiation pulse decays we obtain $I_i = 0$. By solving equation (134) for an initial condition:

$$Q_{bn}\,(0) = \tau_n\,(I_b - I_{c\,(sat)}/B) + I_i\,(1+B)/B$$

we obtain:

$$Q_{bn}\,(t) = \tau_n(I_b - I_{c(sat)}\,/B) + \tau_n I_i\,\frac{(1+B)}{B}\,e^{-1/\tau_n}. \qquad (235)$$

The dissipation time is found from the equation:

$$Q_{bn}t_{nr} = 0 \qquad \text{or} \qquad t_{nr} = \tau_i \ln \frac{I_i\,(1+B)}{I_{c(sat)} - I_b \cdot B}. \qquad (236)$$

At low excitation levels of the non-equilibrium carriers, when their concentration in the base does not exceed the concentration of majority carriers, we have $\tau_n = \tau_b$.

The dissipation time for transistors in which saturation leads to the accumulation of minority carriers (mainly in the collector), may be determined approximately with the aid of equation (236), provided that $I_i = 0$ immediately after irradiation is stopped. A more exact solution of equation (234) for $i_i(t) = I_i\,erfc\,\sqrt{t/\tau_c}$ produces a transcendental equation for t_{nr}:

---

* Without taking into account the peripheral regions

$$\text{erf} \sqrt{t_{nr}/\tau_c} - \frac{2}{\sqrt{\pi}} \sqrt{t_{nr}/\tau_c}\; e^{-t_{nr}/\tau_c} = \qquad (237)$$

$$= \frac{(1+B)\,I_i + B \cdot I_b - I_c\;(sat)}{BI_i}.$$

Beaufoy and Sparks /286/ found that the specific parameters which determine the charge of the equivalent circuit are related to the basic parameters of transistors:

$$\tau_c \equiv \frac{1.22}{\alpha \cdot w_\alpha}; \qquad (238)$$

$$\tau_b = B\tau_c; \qquad (239)$$

$$Q_{be} = Q_{bc} = \frac{q\,p_n w_b A_e}{2} = \tau_c \,|\,I_c\,|\, \exp\,(-qU'_1/kT). \qquad (240)$$

In the above equation α and ω_α are the emitter current transport coefficient and the limiting angular frequency respectively.

The parameter τ_c can also be found by measuring the dependence of the angular frequency ω_t at which $|B| = 1$ /195/, or at which the product $f\,|B_f|$ for the emitter current is described by the equation /14/:

$$1/2\,\pi f\,|\,B_f\,| = \tau_i + \frac{kT}{qI_e}\,(C_1 + C_3). \qquad (241)$$

5. Experimental Data on the Effect of Pulsed Padiation on Semiconductor Devices.

Experimental studies of the effect of pulsed X-ray and γ-radiation on simiconductor devices have been carried out at high dose rates. The equipment included pulsed X-ray devices providing 600 keV radiation pulses lasting several tenths of a μsec and doses of up to several rads (in silicon), pulsed reactors of the "Godiva" type producing γ-radiation pulses (40 to 200 μsec), at maximum dose rates of 10^7 rad / sec and up /275/, and linear electron accelerators producing pulses several tenths of one μsec long, at maximum dose rates up to $3 \cdot 10^{10}$ rad/sec /274/.

Bell-shaped reactor radiation pulses are usually long enough for the establishment of a transient ionization current in semiconductor devices. Therefore, the current pulse envelope almost exactly coincides

251

with the radiation pulse envelope and at each moment of time the current corresponds to the steady-state value at the given radiation dose rate.

Values of the steady-state current in several diodes, measured under irradiation with the maximum pulses of the "Godiva" reactor were compared with the steady-state current, calculated from equation (200) /275/. The pair production rate G was calculated with the aid of the following equation:

$$G = N_{\gamma} \cdot \mu_{lin} \cdot \frac{\overline{E}_c}{\varepsilon} , \qquad (242)$$

where N_{γ} is the gamma-ray density (photons/$(cm^2 \cdot sec)^{-1}$); μ_{lin} is the linear absorption coefficient of γ-radiation by the substance as a result of Compton scattering, cm^{-1}; E_c is the energy of Compton electrons, averaged over the γ-radiation spectrum (in this case 0.223 Mev); ε is the mean energy of formation of an electron-hole pair (3.7 eV for germanium and 4.7 eV for silicon).

The experimental data and calculated values of the currents are listed in Table 13 together with data on the type of the devices, their design and frequency characteristics and the diffusion length of minority carriers. The calculated values of pair production rates G and of the integrated neutron flux of the pulses Φ are also given in the Table.

For diodes, the agreement between experimental and calculated data is usually satisfactory but for transistors the differences are considerable, apparently because of the incorrect determinations of the effective volume of carrier collection (caused by difficulties in the estimation of the contribution of peripheral base regions).

In contrast to pulsed reactors, X-ray devices (which produce very short radiation pulses, i.e., 0.1–0.2 μsec) allow us to investigate the rise and decay of radiation-induced currents. The pulse shape should be as near as possible to the rectangular, otherwise the interpretation of the data and their comparison with calculated results would be rather difficult. This is evident from Figure 102 which shows calculated and experimental current pulses, at the collector junction of 2N1051 n-p-n mesa transistors irradiated with nearly exponential 0.1 μsec X-ray pulses (absorbed dose 0.56 rad). The differences between the amplitudes are apparently due to inaccuracies in the determination of the absorbed dose and of the effective volume of carrier production. It is

Table 13. Comparison of experimental and calculated values of ionization currents in some semiconductor devices.

Type of device	f_α MHz	Diffusion length of carriers, 10^{-3} cm	Surface area of junction 10^{-4} cm²	Integrated flux, 10^{12} neutron/cm²	Rate of pair production 10^{-20} cm³·sec⁻¹	Ionization current, μamp		Ratio of experimental to calculated value
						Calculated	Experimental	
Silicon diode with small junction surface area	—	1.5	3.5	4.2	4.7	130	188	1.45
				6.7	7.4	200	154	0.77
				6.2	6.9	190	347	1.83
				0.55	0.61	17	44.5	2.62
Silicon diode with small junction surface area	—	1.45	0.36	5.0	5.6	14	42	3.00
				8.1	9.1	22	35	1.59
Silicon diode with large junction surface area	—	2.5	6.0	7.3	8.7	630	660	1.05
		5.0						
Alloyed p-n-p germanium transistor	8.0	—	—	7.0	25.0	4900	684	0.14
				7.0	25.0	5100	1200	0.23
				0.35	1.3	250	153	0.61
				0.046	0.17	33	3.6	0.11
Melt-grown n-p-n germanium transistor	8.0	2.1	5.3	7.2	26.0	460	250	0.54
Γ Alloyed p-n-p germanium transistor	10.0	1.8	3.0	7.0	25.0	2200	300	0.14
				0.39	1.5	130	105	0.81
				0.052	0.19	16	9	0.56

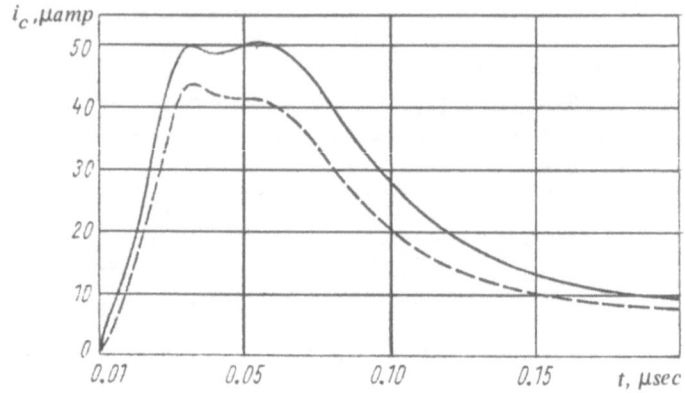

Figure 102. Real (continuous line) and predicted (dash line) ionization current pulses for the collector junction of 1N1051 transistors.

rather difficult to determine the factors responsible for the differences between the pulse shapes; nevertheless the results, as a whole, show good agreement if we take into account the fact that the geometry of diffused mesa transistors (shown in Figure 103) is not unidimensional,

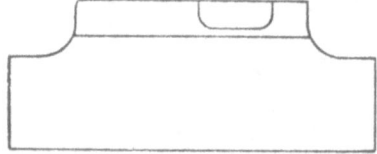

Figure 103. Cross section of a mesa transistor prepared by the double diffusion method.

as was assumed in the derivation of the above equations for the ionization current in *p-n* junctions. The dash line in Figure 102 was obtained by assuming that the entire base region participated in the junction current generation. The error becomes even larger if we assume that the active volume is equal to the product of the collector junction surface area and the distance between the emitter and the collector.

As we mentioned above, in pulsed diodes with infinitesimally short minority carrier lifetimes, the shapes of the ionization current pulses are expected to be similar to those of the radiation pulses and their

254

amplitudes should depend on the reverse voltage. Figure 104 shows oscillograms of current pulses passing through such a diode and of the accelerated electron beams used to irradiate the diode. It is evident that the shapes of the two oscillograms are almost identical. The similar effect of junction voltage on the amplitude of the current in fast-acting 1N696 diodes and on $1/C_{pn}$ (reciprocal of the p-n junction capacitance) indicates (Figure 105) that the instantaneous current component, which is determined solely by the junction width, is predominant in diodes of that type.

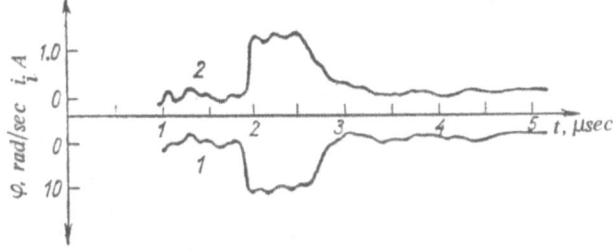

Figure 104. Oscillogram of radiation pulses (1) and of current pulses (2) for a diode with a short minority carrier lifetime. Dose rate $\sim10^{10}$rad/sec; pulse duration 0.6 μsec, amplitude of pulsed current = 1400 milliamperes.

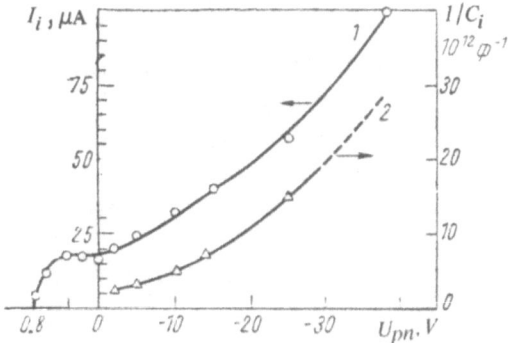

Figure 105. Dependence of the ionization current (1) and of the reciprocal of the junction capacitance (2) on the bias voltage of 1N696 diode. (Duration of pulse 0.1μsec, radiation dose 2.0 rad).

The observed decrease in the current through forward-biased diodes, caused by the appearance of an electric field in the bulk of the material, also agrees well with the function represented by equation (204). If the

255

bulk resistance of the diode base is greater than the differential resistance of the *p-n* junction, the ionization current can further decrease at high forward bias /285/. In this case the voltage drop caused by the ionization current in the bulk of the material would raise the voltage across the junction and consequently the forward current passing through it. As a result, the total current through the diode tends to be constant. The resistance of the bulk and the resistance of the external circuit form, together with the differential resistance of the junction, a self regulating system. Since under those conditions the forward current is determined mainly by the bulk resistance of the diode base, forward ionization currents can be generated at high radiation dose rates, when the conductivity of the material is considerably increased.

Experimental checking of the main results obtained with the equivalent transistor circuit, which is a charge-controlled device, was carried out mainly with melt-grown low-power 2N336 silicon *n-p-n* transistors whose geometry is closest to the above unidimensional model. The sources consisted of X-ray generators producing 0.1 to 0.2 μsec pulses. It should be noted that the shape of such short pulses is of little importance since the parameters of collector current pulses are determined, in the first place, by the time constants of the transistor and by the total excess charge of majority carriers in the base, i.e., by the radiation dose absorbed per pulse.

Figures 106 and 107 show characteristic oscillograms of collector current pulses $i_c(t)$ and of voltages across the base $U_{be}(t)$ recorded during irradiation of such transistors with X-ray pulses /285, 290/.

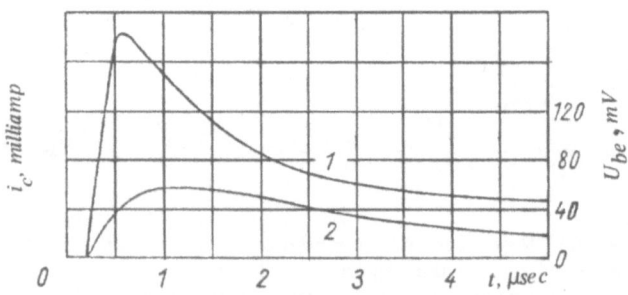

Figure 106. Oscillograms of collector current pulses (1) and of $U_{be}(t)$ voltage pulses (2) generated in a 2N336 transistor by 0.2 μsec X-ray pulses. (The initial collector current was 0.1 mA and the dose was 0.2 rad).

256

Figure 107 shows a collector current pulse, predicted on the basis of transistor specifications or measurements of transistor parameters. Good agreement was observed (see Figure 107).

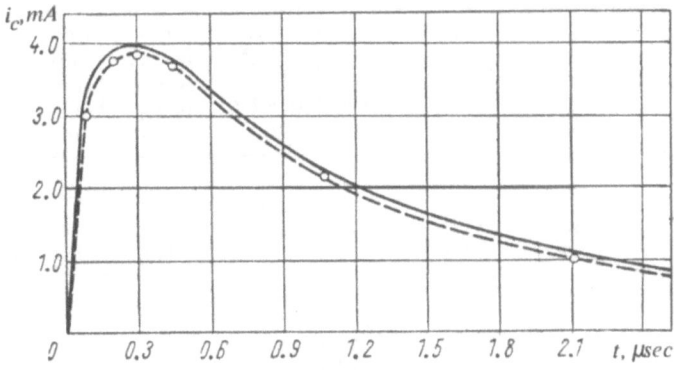

Figure 107. Real (continuous line) and predicted (dash line) collector current pulses in a 2N336 transistor. (Pulse duration 0.1 μsec, radiation dose 0.5 rad (Si))

In both cases the transistors were part of a common-emitter circuit. The collector current pulses were recorded using a small resistor connected in series in the collector circuit. The voltage pulse across the base U_{be} (t) was recorded using an external resistor (R = 100 kohm) which was used to feed into the base the initial current which controlled the operating conditions of the transistor.

Figure 107 shows that, as predicted, both the collector current and the voltage across the base simultaneously increase in the course of 1 μsec; however, the maximum voltage lags somewhat behind the maximum current. Such a delay (whose magnitude is little affected by the collector voltage) is related to the time constant of the base circuit, which is equal to the product of the high transverse resistance of the base (up to 17 kohm) and the external capacitance connected in parallel to the emitter junction.

Irradiation of other types of transistors (alloyed germanium with p-n-p structure) with pulsed X-rays yielded similar results /290/. The characteristics of the collector current pulses and some parameters of the irradiated devices are given in Table 14. Better agreement between the rise times can hardly be expected because of the absence of accurate data on the geometric dimensions of transistors and on the real

257

Table 14. Comparison of collector current pulses for transistors of various types.

Characteristic	Type of device	$f_{\alpha'}$ MHz	$i_{c\,max}$ at $D_\gamma=0.2$ rad, mA	Rise time of current pulse, μsec	Decay time of current pulse μsec	Measured value of $\tau_{b'}$, μ sec
Ge, Alloyed , $p-n-p$ at $U_{c.e}=6.4\,V$	2N1099	0.07	300	7.0	68.0	72.0
	2N174	0.07	300	0.7	47.0	51.0
	2N369	1.0	40	1.7	11.2	9.65
	2N396	8.0	35	0.7	1.4	1.59
	2N525	—	20	1.0	7.0	5.99
	2N269	12.0	15	0.6	0.7	1.13
	2N370	30.0	13	0.6	1.5	1.38
	2N384	100.0	6.5	0.8	0.5	0.56
Si, melt-grown $n-p-n$ at $U_{c.e}=6.0\,V$	2N336 Texas Instruments	15.0	0.15	0.6	1—2	2—3
	2N336 General Electric	15.0	2.0	2.0	4	1.5

lifetimes of the carriers, and because it is impossible to determine the exact influence of the peripheral base regions.

Experimental values of the rise times are of the same order of magnitude as values calculated on the basis of the limiting frequency f_α (Table 15).

We mentioned above that the decay time of current pulses is determined by the lifetime of minority carriers in the base. However, this is true only for devices in which the excess charge of majority carriers is neutralized by minority carriers injected from the emitter. Since for p-n-p devices $n_{excess}=p$, the current decay time, too, is determined by the lifetime of holes in the base (τ_p). For the sake of comparison, the last two columns in Table 14 show the current pulse decay times and the lifetimes of minority carriers measured by a method described in /286/.

In alloyed-transistors good agreement is usually observed between the current decay time and the lifetime of minority carriers, although for low-frequency types there may be certain differences because of effects

Table 15. An evaluation of the current rise time for alloyed transistors with different frequency properties.

f_a, MHz	Transit time in the base, μsec	Calculated lifetime of minority carriers in the base, μ sec	Calculated current rise time, μ sec
0.005	30.0	300.0	30—60
0.01	16.0	300.0	15—40
0.07	2.3	75.0	2—6
0.1	1.6	50.0	1.5—4
0.5	0.3	30.0	0.5—1.0
1.0	0.16	20.0	0.35—0.6
5.0	0.032	5.0	0.3
10.0	0.016	5.0	0.3

associated with the frequency dependence of the diffusion capacitance /290/.

In melt-grown 2N336 transistors a large fraction of the excess charge is accumulated in the barrier layer of the emitter junction, but since this charge is neutralized mainly by ionized impurity atoms (rather than by free minority carriers) the pulsed current decay is not directly determined by the recombination rate of such carriers. Therefore, in such devices the decay of the transient collector current should be slower than in transistors in which the decay is determined by the lifetime of minority carriers; at low currents the current drop should be slower since a large fraction of the excess charge is concentrated in the barrier layer.

Such behavior is observed in the case of 2N336 transistors supplied by the General Electric Company. Figure 108 /290/ shows the experimentally determined decay rate of the collector current pulse $i_c(t)$ and of the voltage drop across the base $U_{be}(t)$ for various initial collector currents i_c^0. The voltage $U_{be}(t)$ decreases slower than the current $i_c(t)$. Such a difference is understandable for transistors of the 2N336 type in which the capacitance of the barrier layer (of the emitter) is higher than the diffusion capacitance, particularly at low collector currents. In addition, at lower initial currents the voltage pulses across the base have higher amplitudes because the diffusion capacitance is proportional to the current and, all other conditions

259

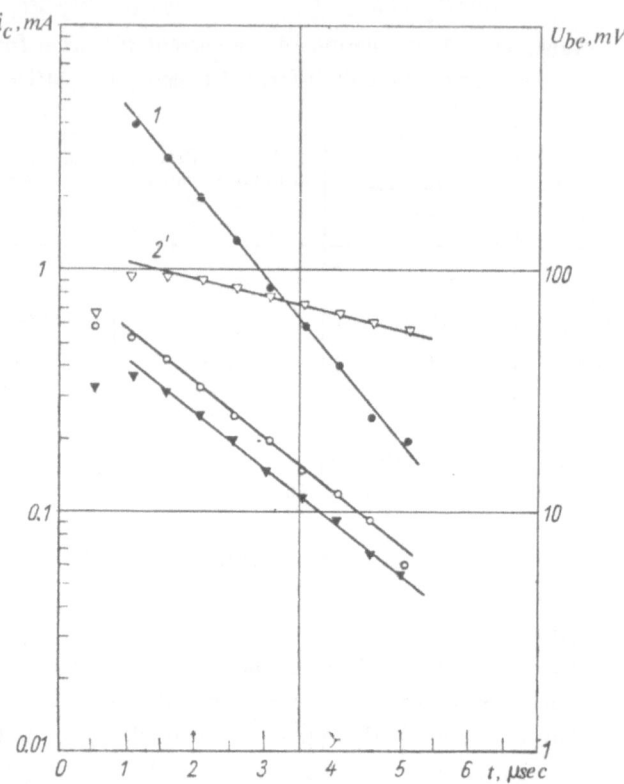

Figure 108. Decay rate of collector current (1, 1') and of the voltage decrease across the base (2. 2') for various values of the initial transistor current i_c^0: 1,2 - 1 mA; 1',2' - 0.01 mA.

being equal, the excess charge accumulated in the nearly constant volume of the barrier layer is inversely proportional to the current.

The dependence of the current pulse amplitude $i_{c\ (max)}$ on the total dose (for radiation pulses with a duration shorter than all time constants of the transistors) (Figure 109) confirms the predicted linear relationship between the absorbed dose, the excess charge and the maximum current for alloy type devices.

For currents smaller than 10 mA, an exponential dependence of the maximum collector current on the dose was expected for 2N336 transistors. However, Figure 109 shows that the dependence is intermediate between linear and exponential. The current increases tenfold when the dose is doubled. The current limit of 100 mA at doses

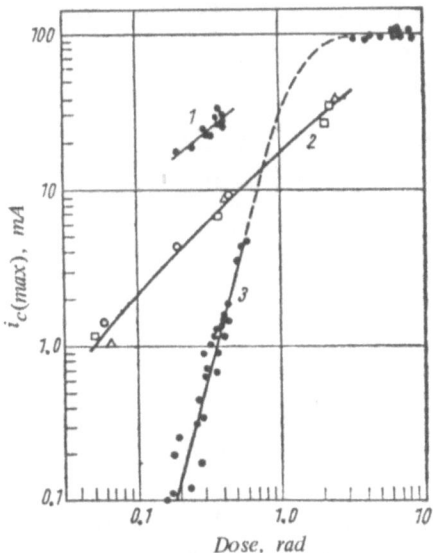

Figure 109. Dependence of the maximum values of the collector current on the radiation dose for various transistors:

1-2N1099, R_{be} varies from 0 to ∞; 2-2N393; Δ-$i_{c_0}^0$ =0.094 mA; O-i_c^0 = 1.04 mA; $\square$-i_c^0 = 9.0 mA; 3 - 2N336, R_{be}= ∞, $i_c^0 <$ 1 μA

above 3 rad is associated with saturation as a result of the relatively high resistance of the bulk of the collector (150 ohm for the above transistor).

At high initial collector currents and absorbed doses (per pulse) multiplication effects take place in 2N336 and other transistors with a high-resistivity collector, because of the appearance of an electric field. As in the case of low currents, here too the dependence of $I_{c(max)}$ on the dose is not linear (Figure 110).

Figure 111 /289/ shows the dependence of the amplitude of collector current pulses on the initial collector current at various absorbed X-ray doses for radiation-resistant alloyed-diffused mesa transistors 2N705 and NS480 (2N760). The shaded area represents the distribution of experimental data for five devices. Each continuous line represents experimental data for one transistor and the dash line represents the same data calculated by an analog computer.

Transistors of the above types, as well as melt-grown 2N336 transistors show a nonlinear dependence of the collector current on the

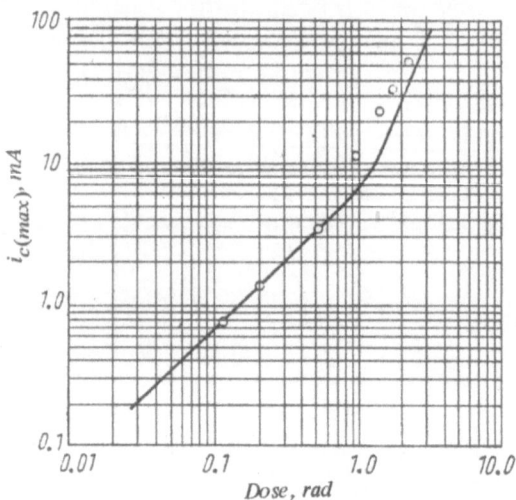

Figure 110. Experimental (points) and calculated (curve) dependence of the maximum collector current on the radiation dose for 2N336 transistors. (The duration of the pulse was 0.1 μsec, and the initial collector current was 5 mA).

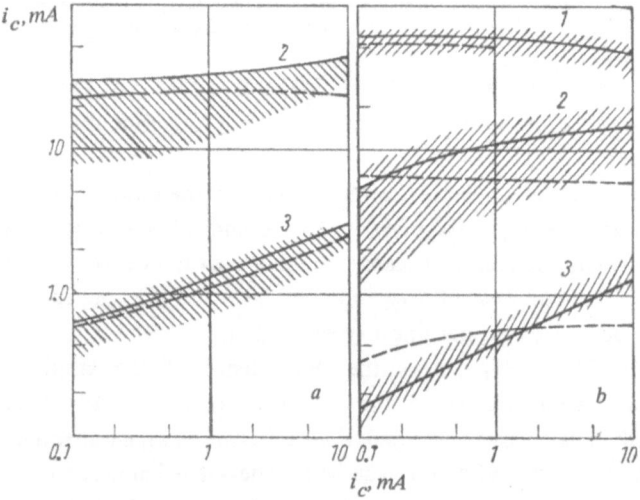

Figure 111. Dependence of the amplitude of collector current pulses on the initial collector current for transistors 2N705 (*a*) and NS480 (2N760) (*b*) at different absorbed doses: 1 - 1.5; 2 - 0.2; 3 - 0.02 rad.

absorbed dose for small initial collector currents at small absorbed doses. As the doses and the initial current increase, this dependence becomes nearly linear.

Despite the scattering of data and the differences between calculated and experimental functions the agreement between the results should be considered satisfactory. The following assumptions were made in the computer programs.

a. τ_c was assumed constant and equal to its value at a collector current of $1\,mA$.

b. τ_b was assumed to be proportional to B at any current.

At first glance it appears that by using equations (238) and (239) and measuring the dependence of B on I_c and of ω_α on I_c it would be possible to determine the dependence of τ_b and τ_c on the collector current. However, this is not so. In the equivalent circuit, τ_b is the effective lifetime of minority carriers in the base, which is proportional to the base current transport factor B, since the latter is a function of lifetime. A plot of the dependence of B on I_c yields a curve with an extremum (the gain increases at first, reaching a maximum and then decreases). The monotonously increasing section of that curve is associated with the increase in effective lifetime. However, the subsequent decrease in B, with increasing I_c, is caused by the decrease in the injection coefficient (not of the lifetime). Therefore, the assumption that τ_b is proportional to B is true only for currents smaller than those at which B reaches a maximum.

The dependence of ω_α on I_c varies in the same way. At relatively large currents the dependence of ω_α on I_c is governed by the dependence of the base width on I_c. At small currents, however, the rapid decrease in ω_α is determined by the time constant $r_e C_1$ rather than by the properties of the base. Therefore, the assumption that τ_c is constant and equal to its value at a collector current of $1\,mA$ is true only for currents close to that value. In reality τ_c may decrease slightly with increasing current. Qualitatively, it could be expected that the dependence of τ_b and τ_c on I_c would cause an increase in the collector ionization current. At lower currents the dependence of τ_c on I_c results in a lower collector ionization current. Thus, the dependence of τ_b and τ_c on the current results in improved agreement between the calculated and experimental values of the collector current (Figure 111).

The data of /274/ show that for most transistors the steady-state collector junction ionization current $I_{i(col)} = qGA_c L_c$ may be calculated, with a relatively small error, from the experimentally determined dissipation time t_p for saturated transistors (Table 112) without knowing the surface area of the junction A_c In that case the error is increased by a factor of two or less.

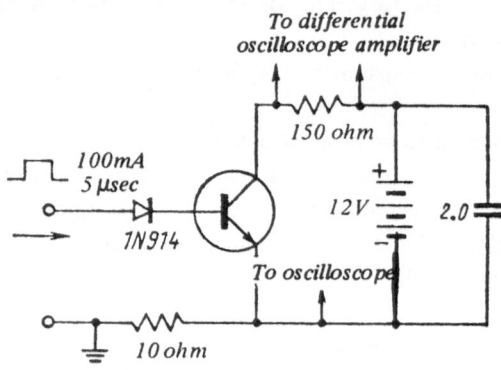

Figure 112. Circuit for measuring the dissipation time t_p (using a pulse of duration t_{pb} = 5 μsec).

Figure 113 shows the dependence of the experimentally measured steady-state collector junction ionization current on $\sqrt{t_p}$ at an absorbed dose rate of $6.6\cdot10^7$ rad/sec for 16 silicon planar and mesa transistors. The dissipation time t_p was measured by feeding into the base 5 μsec current pulses with an amplitude of 100 mA. The saturation collector current $I_{c(sat)}$ was equal to 80 mA.

Since $I_{i(col)}$ is proportional to $\sqrt{D}$ the experimentally determined values of $I_{i(col)}$ for devices with a p-n-p structure can be normalized with the aid of the coefficient $\sqrt{D_n/D_p}$ = 1.6, plotted on the same graph as the data for n-p-n devices and used to derive an empirical equation for the calculation of the ionization current. It is interesting to note that for several types of germanium alloy transistors the values of the collector junction ionization current agree with this curve if the measured $I_{i(sat)}$ values are normalized to account for differences between the rates of carrier production in silicon and germanium as well as for differences between the diffusion coefficients of minority carriers in the bases of germanium transistors and in the collectors of silicon n-p-n transistors.

264

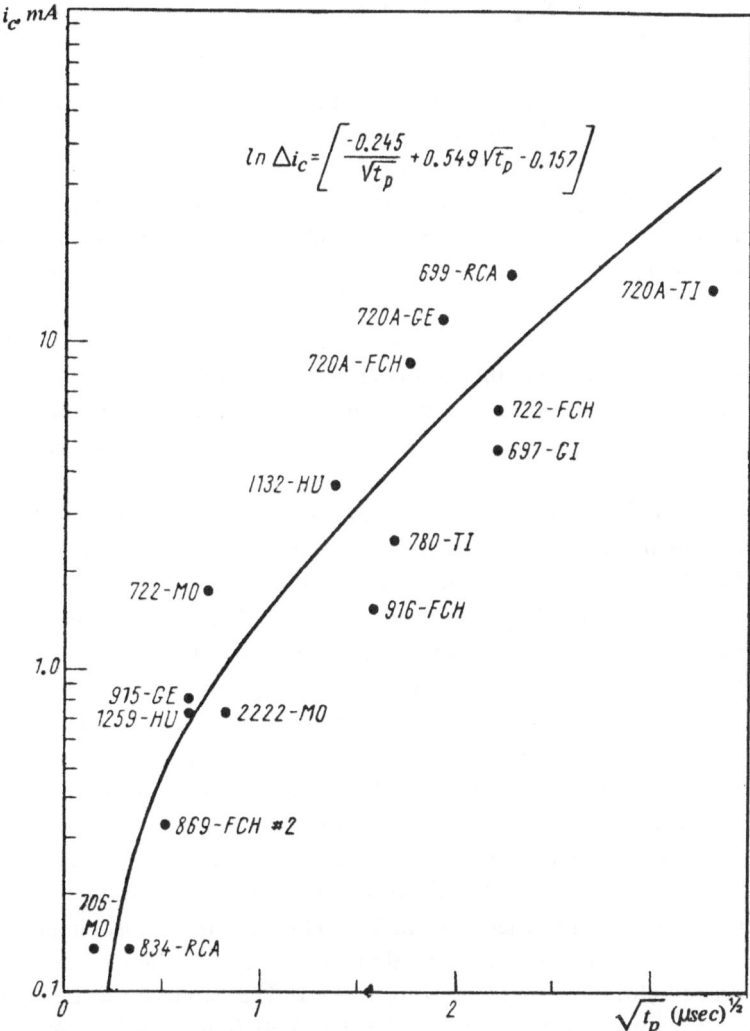

$$\ln \Delta i_c = \left[\frac{-0.245}{\sqrt{t_p}} + 0.549\sqrt{t_p} - 0.157\right]$$

Figure 113. Corelation between the ionization current of the collector junctions of silicon planar and mesa transistors of various types and $\sqrt{t_p}$ at $\varphi = 6.6 \cdot 10^7$ rad/sec.

It was also shown that if the ionization current is proportional to the radiation dose rate the steady state current can be calculated from the experimentally determined dissipation time (before irradiation), using the following equation:

$$I_{i(col)} = \frac{K_\gamma \varphi_\gamma}{6.6 \cdot 10^7} \cdot 10^{\left[-\frac{0.245}{\sqrt{t_p}} + 0.55 \sqrt{t_p} - 0.157 \right]}, \tag{243}$$

($I_{i(col)}$ is in mA; φ_γ in rad/sec and t_p in μsec). K_γ depends on the material and on the structure of transistors and its values are as follows: $K_\gamma = 1.0$ for silicon n-p-n planar and mesa transistors; $K_\gamma = 6.8$ for germanium p-n-p alloy transistors; $K_\gamma = 10.1$ for silicon n-p-n alloy transistors; $K_\gamma = 1.6$ for silicon p-n-p planar on mesa transistors. The value of t_p should be measured at $I_{c(sat)} = 80\ mA$, $I_b = 100\ mA$, and $t_{pb} = 5\ \mu$sec.

Thus, pulsed irradiation of transistors and of diodes is accompanied by transient processes that can considerably impair the performance of electronic devices. Such transient processes cause redistribution and annihilation of structural defects as well as relaxation of electron excitations.

The redistribution and annihilation processes may affect the performance of such devices for tens and hundreds of seconds and even after that the initial parameters may never be fully recovered.

Studies of the kinetics of defect annealing began only recently and many phenomena remain unclear. It is therefore difficult to predict the magnitude and duration of transient changes in the parameters of devices, caused by the development and annealing of structural defects.

Processes caused by electron excitation or ionization in semiconductor devices have been investigated in greater detail. It was found that such processes may be represented by equations for equivalent circuits in which the diodes and the transistors are charge-controlled. However, those equations are valid only within the limits of the assumptions that are made during their derivation. One of the most important assumptions involves a condition for a low excitation level of non-equilibrium carriers, i.e. $\Delta n/n_0 \leqslant 1$. That condition corresponds to absorbed dose rates (ionizing radiation) not exceeding $10^7 - 10^8$ rad/sec. Empirical expressions have been obtained for higher absorbed dose rates ($10^9 - 10^{11}$ rad/sec). Therefore it is necessary to carry out further investigations on the operation of semiconductor devices at high excitation levels of non-equilibrium charge carriers.

BIBLIOGRAPHY

1. Proceedings of the 7th International Conf. on the Physics of Semiconductors. Part 3. Radiation Damage in Semiconductors. Dunod, Paris, 1965.

2. Proceedings of the 8th International Conf. on the Physics of Semiconductors. Tokyo, 1966;

3. Actions des rayonnements sur les composants a semiconducteurs. Toulouse, March 1967, France.

4. Proceedings of the Confer. on Radiation Effects in Semiconductors. *J. Appl. Phys.*, **30**, No. 8 (1959).

5. Proceedings of the International Conf. on Crystal Lattice Defects. Kyoto, Japan, 1962; *J. Phys. Soc. Japan*, **18**, 3 (1963).

6. J. Corbett, Electron Rad. Damage in Semiconductors and Metals. *Progress in Solid State Physics*, 1967.

7. A.F. Joffe. Semiconductor Physics (in Russian) Izd-vo AN SSSR, Moscow, 1957.

8. R. Smith. Semiconductors (in Russian) Izd-vo Inostr. Lit. (Moscow) 1967.

9. Experimental Nuclear Physics (in Russian). Edited by E. Segre. Vol. 1. Izd-vo Inostr. Lit. (Moscow) 1955.

10. G. Dearnaley and D. Northrop. Semiconductor Counters of Nuclear Radiation (in Russian) Mir, Moscow 1966.

11. F. Seitz *et al. Solid State Phys.*, **2**, 307, No. 4 (1956).

12. J. Dienes and J. Viniard. Radiation Effects in Solids (in Russian) Izd-vo Inostr. Lit., Moscow 1960.

13. V. S. Vavilov. Effects of Radiation on Semiconductors (in Russian) Fizmatgiz, Moscow 1963.

14. J. Crawford, In: Actions Cliniques et Biologiques des Radiations. Ed. by M. Haissinsky, Masson, Paris, 1964.

15. R. Kaiser, *Z. angew. Phys.*, **19**, 3, 270 (1955).

16. W.A. McKinley, H. Feshbach, *Phys. Rev.*, **74**, 1759 (1948).

17. J. Kahn, *J. Appl. Phys.*, **30**, 1310 (1959).

18. G. Kinchin, R. Rease, *Repts. Progr. Phys.*, **18**, (1955); *Uspekhi Fiz. Nauk* **60**, 590, (1956).

19. H. Smith, *Philos. Trans. Roy. Soc.*, **A241**, 105 (1948).

20. W. Kohn, *Phys. Rev.*, **94**, 1409 (1954).

21. W. Brown, W. Augustinak, *J. Appl. Phys.*, **30**, 1300 (1959).

22. N.S. Smirnov, P. Ya. Grazinov, *Fizika Tverdogo Tela*, **1**, 1376 (1959).

23. W. Snyder, J. Neufeld, *Phys. Rev.*, **97**, 1636 (1955).

24. P.I. Gregoriev and A.I. Fedorenko, *Uspekhi Fiz. Nauk* **83**, 385 (1955).

25. J. Brinkman, *J. Appl. Phys.*, **25**, 961 (1954).

26. L.S. Smirnow, *Fizika Tverdogo Tela*, **2**, 1669 (1960).

27. D. Wiegand, R. Smoluchowsky, Actions Cliniques et Biologiques des Radiations. Ed. by M. Haissinsky Masson, Paris, 1964.

28. J. Varley, *Nature*, **174**, 886 (1954).

29. J. Varley, *J. Nucl. Energy*, **1**, 130 (1954).

30. R. Howard *et al. Phys. Rev.*, **122**, 1406 (1961).

31. J. Varley, *J. Phys. Chem. Solids*, **23**, 985 (1962).

32. R. Howard, R. Smoluchowsky, *Phys. Rev.*, **116**, 314 (1959).

33. D. Dexter, *Ibid.*, **118**, 934 (1960).

34. S.V. Starodubtsev and A.E. Kiv, Electronic Processes in High Energy Chemistry (in Russian) Nauka, Moscow 1965, p. 278.

35. Yu. A. Kurskii, *Fizika Tverdogo Tela*, **6**, 1485 (1964).

36. H. James, K. Lark-Horowitz, *Z. Phys. Chem.*, **198**, 107 (1951).

37. J. Lofersky, *et al. Phys. Rev.*, **111**, 432 (1958).

38. R. Bänerlein, *Z. Phys.*, **176**, 498 (1963).

39. B. Kulp, *et al. Phys. Rev.*, **129**, 2422 (1963).

40. B. Kulp, *Ibid.*, **125**, 1865 (1962).

41. V.S. Vavilov, and A.F. Plotnikov, *Fizika Tverdogo Tela*, **3**, 8 (1961).

42. V.S. Vavilov, A.F. Plotnikov, V.D. Tkachev, *Fizika Tverdogo Tela*, **4**, 3446, 3575 (1962); **5**, 3188 (1963).

43. I.P. Akimchenko, V.S. Vavilov, and A.F. Plotnikov, *Fizika Tverdogo Tela*, **5**, 1417 (1963).

44. A.F. Plotnikov, V.S. Vavilov, and V.D. Kopilovski, *Pribory i Tekhnika Eksperimenta* No. 3, 183 (1962).

45. T. Moss, Optical Properties of Semiconductors, Izd-vo Inostr. Lit. 1960, (in Russian).

46. V.S. Vavilov, E.N. Lotkova and A.F. Plotnikov in: Photocon-

ductivity in Solids. Pergamon Press, London, 1962, p. 32.

47. P. Byub, Photoconductivity in Solids (in Russian) Izd-vo Inostr. Lit., Moscow 1963.

48. E.I. Adirovitch, *Fizika Tverdogo Tela*, **2**, 2248 (1960).

49. S.M. Rivkin *et al. Ibid.*, **3**, 252 (1961).

50. V.S. Vavilov, Proceedings of the 7th International Conf. on the Physics of Semiconductors. Paris, 1964; Rad. Damage in Semiconductors. Dunod, Paris, 1965, p. 115; *Uspekhi Fiz. Nauk*, **84**, 431 (1964).

51. A.F. Plotnikov, V.S. Vavilov, and L.S. Smirnov, *Fizika Tverdogo Tela*, **3**, 3254 (1961).

52. S.M. Rivkin, Photoelectric Phenomena in Semiconductors. Moscow, Fizmatgiz. 1963.

53. Collection of articles edited by Penin N.A. Izd-vo Inostr. Lit. Moscow 1962, (in Russian).

54. J. Ludwig, G. Woodberry, Electron Spin Resonance in Semi-conductors (in Russian) Mir, Moscow 1964.

55. G. Watkins, Proceedings of the 7th International Conf. on the Physics of Semiconductors. Paris, 1964, Part 3; Rad. Damage in Semiconductors..Dunod, Paris, 1965, p. 97.

56. G. Bemski, *J. Appl. Phys.*, **30**, 1195 (1959).

57. V.S. Vavilov, S.I. Vintovkin, A.R. Retovich, A. F. Plotnikov and A.A. Sokolova, *Fizika Tverdogo Tela*, **7**, 502 (1965).

58. J. Corbett, G. Watkins, *Phys. Rev. Letters*, **7**, 314 (1961).

59. G. Watkins, *J. Phys. Soc. Japan*, **18**, Suppl. 2, 22 (1963).

60. H. Fan, A. Ramdas, *J. Appl. Phys.*, **30**, 1127 (1959).

61. J. MacKay, Proceedings of the 7th International Conf. on the Physics of Semiconductors. Paris, 1964, Part 3; Rad. Damage in Semiconductors. Dunod, Paris, 1965, p.11.

62. J. Duncan, *Ibid.*, p. 135.

63. B. Kulp, *Ibid.*, p. 131.

64. E. Sonder, *et al. J. App.. Phys.*, **30**, 1186 (1959).

65. B.I. Boltax, Diffusion in Semiconductors (in Russian) Moscow, Fizmatgiz, 1961, p. 264.

66. G. Bemsky *et al. J. Phys. Chem. Solids*, **24**, 1 (1963).

67. E. Pell, *J. Appl. Phys.*, **32**, 1048 (1961), 113.

68. V.S. Vavilov *et al.* *Fizika Tverdogo Tela,* **4,** 1128, 3373 (1962); *J. Phys. Soc. Japan,* **18**, 3, 136 (1963). I.V. Kryukova, and V.S. Vavilov, *Fizika i Tekhnika Poluprovodnikov,* **2**, 1312 (1968).

69. G. Watkins, *et al. J. Appl. Phys.,* **30**, 1198 (1959).

70. D. Hill, *Phys. Rev.,* **119**, 1222 (1960).

71. G. Bemsky *et al. Ibid.,* **108**, 645 (1957).

72. K.V. Lark-Horovitz Semiconductor Materials (in Russian). Izd-vo Inostr. Lit. 1954, p. 62.

73. O.L. Curtis, J. Crawford, *Phys. Rev.,* **124**, 1731 (1961); **126**, 1342 (1962).

74. R. Logan *Ibid.,* **101**, 1455 (1956).

75. V.V. Antonov-Romanovskii, *Optika i Spektroskopiya* **3**, 592 (1957).

76. W. Brown *et al. Phys. Rev.,* **92**, 591(1953).

77. T. Waite *Ibid.,* **107**, 471 (1957).

78. A.V. Spitsin, L.S. Smirnov, *Fizika Tverdogo Tela* **4**, 3456 (1962).

79. S. Ishino *et al. J. Phys. Chem. Solids,* **24**, 1033 (1963).

80. E.Blount, *J. Appl. Phys.* **30**, 1218 (1959).

81. N.A. Vitovskii, T.V. Mashovets and S.M. Ryvkin, *Fizika Tverdogo Tela,* **5**, 1833 (1963).

82. J. Cleland, *J. Appl. Phys., Letters,* **3**, 113 (1963).

83. F. Eisen, *Phys. Rev.,* **123**, 736 (1961).

84. F. Eisen, *Bull. Amer. Phys. Soc.,* **9**, 290 (1964).

85. J. Kortright, J. MacKey, *Ibid.,* **3**, 142 (1958).

86. B. Gossick, *J. Appl. Phys.,* **30**, 1204 (1959).

87. J. Cleland, *et al. Phys. Rev.,* **99**, 1170 (1955).

88. H. Juretschke, *et al. J. Appl. Phys.,* **17**, 838 (1956).

89. J. Cleland, J. Crawford, Proceedings of the International Conf. on Semiconductors. Prague, 1960; Academic Press, No. 4, 299 (1961).

90. N. Van Dong *et al. Trans. Faraday Soc.,* **57**, 1968 (1961).

91. H. Stein, *J. Appl. Phys.,* **31**, 1301 (1960).

92. W. Closser, *Ibid.,* **31**, 1693 (1960).

93. B. Gossick, *Ibid.,* **31**, 745 (1960).

94. F. Vook, Proceedings of the 7th International Conf. on

Physics of Semiconductors. Paris, 1964, Part 3; Rad. Damage in Semiconductors. Dunod, Paris, 1965, p. 51; V.S. Vavilov *et al.* Radiation Effects in Semiconductors, N.Y. Plenum Press, 1968, p. 346.

95. R. Simmons *et al. J. Appl. Phys.*, **30**, 1249 (1959).

96. J. Eshelby, *Ibid.*, **15**, 255 (1964).

97. F. Vook, *Phys. Rev.*, **125**, 855 (1962).

98. F. Vook, *J. Phys. Sco., Japan*, **18**, 2, 190 (1963).

99. R. Swalin, *J. Phys. Chem. Solids*, **18**, 290 (1961).

100. S. Kurnick, *J. Chem. Phys.*, **20**, 219 (1952).

101. G. Pearson *et al.* Proceedings of the 7th International Conf. on the Physics of Semiconductors. Paris, 1964, Part 3; Rad. Damage in Semiconductors. Dunod, Paris, 1965, p. 197.

102. F. Vook, *et al. Phys. Rev.*, **113**, 62 (1959).

103. C. Pierce *et al., J. Appl. Phys.*, **34**, 1946 (1963).

104. M. Wittels, *Bull. Amer. Phys. Soc.*, **5**, 375 (1960).

105. F. Fujitax *et al. J. Phys. Soc. Japan*, **13**, 1063 (1953).

106. U. Gonser *et al. Phys. Rev.*, **109**, 663 (1958).

107. D. Keitman, *et al. Ibid.*, **108**, 901 (1957).

108. J. Parsons *et al. J. Appl. Phys. Letters*, **1**, 57 (1962).

109. M. Bertolotti, *et al. Nuovo Cimento*, **29**, 1200 (1963).

110. R. Berman, *Proc. Roy. Soc.*, **A208**, 90 (1951).

111. F. Vook, *Bull. Amer. Phys. Soc.*, **8**, 437 (1963); **8**, 209 (1963); **9**, 289 (1964).

112. J. Ziman, *Canad. J. Phys.*, **34**, 1256 (1956).

113. L. Aukerman, *Phys. Rev.*, **127**, 1576 (1962).

114. G. Arnold *et al. Bull. Amer. Phys. Soc.*, **9**, 220 (1964).

115. N. Van Dong *et al. Compt. Rend.*, **236**, 1722 (1963).

116. M. Vandeiver, H. Albany, *Ibid.*, **257**, 1252 (1963).

117. M.M. Bredov, N.M. Okuneva, *Dokl. AN SSSR*, **113**, 795 (1957); M.M. Bredov, L.V. Lepilin, *et al. Fizika Tverdogo Tela*, **3**, 267 (1961); M.M. Bredov, A.B. Nuromskii, *Ibid.*, **4**, 562 (1962).

118. R.S. Ohl, *Bell System Techn. J.*, **31**, 104 (1952).

119. K. Nielsen, Electromagnetically Enriched Isotopes and Mass-Spectrometry. London, 1958, p. 68.

120. V.M. Gusev, V.V. Titov, M.I. Guseva, and V.I. Kurinnyi, *Fizika Tverdogo Tela*, **7**, 2077 (1965).

121. V.M. Gusev *et al. Ibid.*, **8**, 1708 (1966).

122. P. Baruch *et al. J. Phys. Soc. Japan*, **18**, 3, 251 (1963); J. Pfister, Dissertation. Ecole Normale Superience. Paris, France, 1964.

123. R.P. Ferber, *IEEE Trans. Nucl. Sci.*, **NS−10**, 15 (1963).

124. P.V. Pavlov, E.I. Zorin,, D.I. Tetelbaum and Yu. S. Popov, *Fizika Tverdogo Tela*, **6**, 3222 (1964).

125. G. Dearnaley, Report on International Symposium on Radiation Effects in Semiconductors. Toulouse, France, 1967.

126. P. Pavlov, *et al.* Report to the International Symposium on the Radiation Effect on Semiconductors, Toulouse, France, 1967.

127. V.S. Vavilov, M.I. Guseva, E.A. Konorova, V.V. Krasnopevtsev, V.F. Sergienko and V.V. Titov, *Fizika Tverdogo Tela*, **9**, 1964 (1967).

128. Ion Implantation. Techn. Information No.8, AERE, Harwell, England, 1967.

129. J. Stark, *Z. Phys.*, **13**, 585 (1912).

130. M. Robinson, O.Oen, *J. Appl. Phys. Letters*, **2**, 30 (1963).

131. G. Pierchy *et al. Phys. Rev. Letters*, **10**, 399 (1963).

132. R. Nelson, M. Thompson, *Philips Mag.*, **11**, No. 2, 176 (1966).

133. J. Lindhard, *Phys. Letters*, **12**, 126 (1964).

134. A.F. Tulinov, *Uspekhi Fiz. Nauk*, 585 (1965).

135. E. Bogh *et al. Phys. Letters*, **12**, 129 (1964).

136. G. Dearnaley, *IEEE Trans. Nucl. Sci.* **NS-11**, 249 (1964).

137. F. Eisen, *Bull. Amer. Phys. Soc.*, **11**, 176 (1966).

138. R. Nelson *et al. Phys. Letters*, **23**, 18 (1966).

139. C. Moak, *et al. Bull. Amer. Phys. Soc.*, **11**, 101 (1966).

140. E. Kornelson *et al. Phys. Rev.*, **136**, 1849 (1964).

141. R. Bower *et al. J. Appl. Phys. Letters*, **9**, 203 (1966).

142. J. Davies, *et al. Canad. J. Phys.*, **44**, 1631 (1966).

143. N.P. Kekelidre, Transactions of the IAEA Symposium on the Physical Effects of Nuclear Transformations, Prague, 1960.

144. J. Messier, Y. Lecollere, *IEEE Trans. Nucl. Sci.*, **NS−11**, No. 3 276 (1964).

145. H.Schweinler, *J. Appl. Phys.*, **30**, 1125 (1959).

146. M.V. Chukichev, N.S. Vavilov, *Fizika Tverdogo Tela*, **3**, 1522 (1961).

147. L.V. Groshev, *et al.* Atlas of γ-Ray Spectra from Radiation Capture of Thermal Neutrons. (In Russian) Moscow, Atomizdat, 1958.

148. L.K. Vodop'yanov, N.I. Kurdiani, *Fizika Tverdogo Tela,* **8**, 72, (1966).

149. V.L. Bonch-Bruevich, *Ibid.,* **4**, 2660 (1962); **5**, 1852 (1963).

150. L.V. Keldysh, G.M. Proshko, *Ibid.,* **5**, 3378 (1963).

151. N. Bohr, Transport of Nuclear Particles Through Matter (in Russian) Moscow, Izd-vo Inostr. Lit., 1951.

152. H. Fermi, Nuclear Physics (in Russian) Izd-vo Inostr. Lit., 1951.

153. J. Marion, Nuclear Data Tables, N.Y., Part 3, 1960.

154. C. Moller, *Ann. Phys.,* **14**, 531 (1932).

155. H. Bethe, W. Heitler, *Proc. Roy. Soc.,* **A146**, 83 (1934).

156. F. Seitz, *Disc. Faraday Soc.,* **5**. 271 (1949).

157. Characteristic Energy Losses of Electrons, (in Russian) Collection of Articles, Edited by A.P. Shul'man, Izd-vo Inostr. Lit., 1959.

158. V. Heitler, Quantum Theory of Radiation (in Russian) Moscow, Izd-vo Inostr. Lit., 1956.

159. R. Hofstadter, *Nucleonics,* No. 4,2;No. 5, 29 (1949); *Uspekhi Fiz. Nauk,* **39**, 462 (1949).

160. S.M. Ryvikin, O.A. Matveev, and N.B. Strokan, Semiconductor Counters of Charged Particles, (in Russian) Leningrad, Znanie, 1963.

161. K. MacKay, *Phys. Rev.,* **84**, 829 (1951).

162. V.S. Vavilov, L.S. Smirnov and V.M. Patskevich, *Dokl. AN SSSR,* **112**, 1020 (1957).

163. J. Drahokoupil *et al. Czechosl. J. Phys.,* **7**, 521 (1957).

164. M.V. Chukichev, V.S. Vavilov, *Fizika Tverdogo Tela,* **3**, 935 (1961).

165. G.P. Golubev, V.S. Vavilov and V.A. Egorov, *Ibid.,* **7**, 3000 (1965).

166. C. Klein, Proceedings of the 8th International Conf. on the Physics of Semiconductors, Kyoto, 1966; printed in *Phys. Soc. Japan,* p. 307.

167. H. Philipp, H. Ehrenreich, *Phys. Rev.,* **129**, 1550 (1963).

168. J. Touc, A. Abraham, *Czechosl. J. Phys.,* **9**, 95 (1959).

169. M. Halbert, J. Blankenship, *Nucl. Instrum. and Methods,* **8**, 106 (1960); J. MacKenzie, D. Bromley, *Phys. Rev. Letters,* **2**, 303

(1959); K. MacKay, K. MacAfee, *Phys. Rev.*, **91**, 1079 (1953).

170. A. Zareba, In: Proc. of the International Conf. on the Physics of Semiconductors. Prague, 1960. Prague Academ. Press, 1961, p. 476; V.S. Vavilov *et al. Dokl. AN SSSR*, **93**, 246 (1954); **112**, 93 (1957).

171. S. Koc, *Czechosl. J. Phys.*, **7**, 91 (1957); J. Drahokoupil *et al. Ibid.*, **7**, 57 (1957).

172. F. Emery, T. Rabson, *Phys. Rev.*, **140**, A2089 (1965); E. Baldinger *et al. Z. Andrew. Math. und Phys.*, **15**, 90 (1964); **131**,134 (1964).

173. C. Bussola *et al. Phys. Rev.*, **136**, A1756 (1964); V.M. Patskevich, V.S. Vavilov, and L.S. Smirnov, *Zh. Eksperim, i Teor. Fiz.*, **6**, 619 (1958).

174. L. Koch, *et al. Compt.Rend. Acad. Sci.*, **250**, 74 (1961).

175. J. Pfister, *Z. Naturforsch.*, **12a**, 217 (1957).

176. B. Goldstein, *J. Appl. Phys.*, **36**, 3853 (1965).

177. P. Van Heerden, *Phys. Rev.*, **106**, 468 (1957).

178. F. Lappe, *Z. Physik*, **154**, 267 (1959).

179. F. Lappe, *J. Phys. Chem. Solids*, **20**, 173 (1961).

180. P. Kennedy, *Proc. Roy. Soc.*, **A253**, 337 (1959).

181. W. Shockley, *Solid State Electronics*, **2**, 35 (1961); *Uspekhi Fiz. Nauk*, **77**, 161 (1962).

182. Yu.M. Popov, *Trudy Fiz. In-ta AN SSSR* **23**, 64 (1963).

183. D. Prines, Elementary Excitations in Solids, N.Y. Benjamin, 1963, Chapt.4.

184. D.S. Peck, *et al. Nature*, No. **4895**, 741 (1963).

185. V. Riddle, New Semiconductor Device Circuits (in Russian) Edited by A.A. Sokolov, Izd-vo Inostr. Lit., 1961, p. 9.

186. W.V. Behrens, *et al. Proc. I.R.E.*, **46**, No. 3, 601 (1958).

187. J.W. Easley, *Nucleonics*, **20**, No. 7, 51 (1962).

188. G.C. Messenger, *IEEE Trans. Nucl. Sci.*, **NS—12**, No. 2, 53 (1965).

189. J.W. Easley, *et al. J. Appl. Phys.*, **31**, 1024 (1960).

190. P.O. Lauritzen, *et al. IEEE Trans. Nucl. Sci.*, **NS—11**, No. 5, 39 (1964).

191. E. Baldinger, *et al. Solid State Electronics*, **9**, No. 3, 287(1966).

192. A.V. Krasilov, *et al.* Methods of Semiconductor Calculation (in Russian) Moscow, Energiya, 1964.

193. I.J. Sparkes, *et al. Proc. I.R.E.*, **45**, No. 12, 182 (1957).

194. I.P. Stepanenko, Theoretical Principles of Transistors and Transistor Circuits, (in Russian) Moscow, Gosenergoizdat, 1963.

195. Lindmeier *et al. Trudy In-ta Radioinzhinerov,* **No. 2**, 228 (1962); *Proc. I.R.E.*, **50**, No. 2.

196. N.A. Ukhin, Electrophysical, Dosimetric and Radiometric Equipment (in Russian) Atomizdat, Moscow 1962.

197. J.A. Hood, *Semicond. Prod. and Solid-State Techn.,* **8**, No.1, 13-16 (1965).

198. C. Sah, *et al. Proc. I.R.E.*, **45**, No. 9, 1228 (1957).

199. H. Bilger, *Helv. Phys. Acta,* **XXXIV**, 909 (1961).

200. C. Sah, *IRE Trans. Electron Devices,* **ED-9**, No. 1, 92 (1962).

201. C.A. Goben, *IEEE Trans. Nucl. Sci.,* **NS-12**, No. 5, 134 (1965).

202. J. A. Beicker, Report on International Simposium on Radiation Effects in Semiconductors. Toulouse, France, March 1967.

203. J.W. Easley, In: Proceedings of the 7th International Conf. on the Physics of Semiconductors. Part 3, Paris, 1965, p.341.

204. R.W. Beck, *et al. J. Appl. Phys.,* **30**, No.9, 1437 (1959).

205. O.L. Curtis, *et al. Ibid.,* **28**, No. 10, 1161 (1957).

206. O.L. Curtis, *et al. Ibid.,* **29**, No. 12, 1722 (1958).

207. V.S. Vavilov, *et al. Zh. Eksperim. i Teor. Fiz* **32**, 702 (1957); *Ibid.,* **34**, No. 2 (1958).

208. S.M. Ryvkin, *et al. Fizika Tverdogo Tela,* **3**, 3211 (1961).

209, O.L. Curtis, *et al.* In: Proceedings of the 7th International Conf. on the Physics of Semiconductors. Part 3. Dunod, Paris, 1965, p. 143.

210. N.A. Vitovskii, *et al. Fizika Tverdogo Tela,* **1**, 1381 (1959).

211. K. Matsuura, *et al. J. Phys. Soc. Japan,* **16**, 339 (1961).

212. B.G. Streetman, *J. Appl. Phys.,* **37**, 3145 (1966).

213. R.F. Konopleva, *et al. Fizika Tverdogo Tela,* **6**, 1063 (1964).

214. A.N. Sevchenko, *et al. Dokl. AN SSSR,* **169**, 562 (1966).

215. W. Shockley, *et al.* Semiconductor Electronic Devices (in Russian) Edited by A.V. Rzhanov, Moscow, Izd-vo Inostr. Lit., 120, 1953.

216. J.W. Cleland, *et al. Phys. Rev.,* **83**, 312 (1951).

217. J. Cleland, *et al. Ibid.*, **98**, 1742 (1955).

218. D. Binder, *Ibid.*, **122**, 1147 (1961).

219. A.D. Kantz, *J. Appl. Phys.* **34**, 1944 (1963).

220. G.C. Messenger *et al. Proc. I.R.E.*, **46** (1955).

221. H.C. Paxton, *Nucleonics*, **13**, No.10, 48 (1955).

222. L. Taylor, *IRE Trans. Nucl. Sci.*, **No. 5**, No. 1 280 (1962).

223. A. Walters, In: Proceedings of 2nd Conf. on Nuclear Radiation Effects on Semiconductors Devices. Materials and Circuits. N.Y., Cowan Publishing Corp., 1959, p. 35.

224. R.F. Konopleva, *et al.* Solid State Physics, (in Russian) Part 2, Leningrad Izd.AN SSSR, 1959,11.

225. O.L. Curtis, *et al. J. Appl. Phys.*, **30**, 1174 (1959).

226. L.G. Sharendo, *et al. Fizika Tverdogo Tela*, **4**, 2137 (1962).

227. O.L. Curtis, *et al. J. Appl. Phys.*, **36**, 2094 (1965).

228. A.V. Spitsin, Report to the International Conf. on Radiation Effects in Semiconductors. Toulouse, France, March 1967.

229. M. Bertolotti, *et al. Nuovo Cimento*, *XX*, **No.3**, 438 (1964).

230. M. Hirata *et al. J. Appl. Phys.*, **37**, 1867 (1966).

231. R. Glaenzer, *et al. Ibid.*, **36**, 2197 (1965).

232. G. Wertheim, *Phys. Rev.*, **105**, 2197 (1957).

233. C. Klein, *J. Appl. Phys.*, **30**, 1222 (1959).

234. G. Galkin, *Fizika Tverdogo Tela*, **3**, 630 (1961).

235. G. Galkin, *et al. Ibid.*, 2, 2018 (1960).

236. J. Merlo-Flores *et al.* In: Proceedings of the 7th International Conf. on Semiconductors. Dunod. Paris. 1965, p. 155.

237. E. Sonder *et al. J. Appl. Phys.*, **31**, 1279 (1960).

238. E. Sonder *et al. Ibid.*, **34**, 3296 (1963).

239. E. Sonder *et al. Ibid.*, **36**, 1811 (1965).

240. V.S. Vavilov, *et al. Fizika Tverdogo Tela*, **4**, 1969 (1962).

241. H. Stein, *J. Appl.,Phys.*, **37**, 3382 (1966).

242. O.L. Curtis, *IEEE Trans. Nucl. Sci.*, **NS-13**, No. 6, 33 (1966).

243. G.C. Messenger, In:: Report on International Symposium on Radiation Effects in Semiconductors. Toulouse, March 1967.

244. R. Brown, In: Report on International Conf. on Radiation Effects in Semiconductors. Toulouse, March 1967.

245. U. Cocca, *et al.* In: Radiation Effects in Electronics. ASTM Special Technical Publication, No. 384, 1965, p; 149.

246. C.D. Taulbee, *et al.* In: Radiation Effects in Electronics. ASTM Special Technical Publication, No. 384, 1965, p. 121.

247. R.R. Blair, *IEEE Trans. Nucl. Sci.,* **NS–10**, No. 5 35 (1963).

248. M.M. Weiss *et al. Ibid.,* **NS–10**, No. 5, 28 (1963).

249. G. Brucker. Report on International Conf. on Radiation Effects in Semiconductors. Toulouse, March 1967.

250. M. Frank, *et al.* In: Radiation Effects in Electronics. ASTM Special Technical Publication, No. 384, 1965, p. 173.

251. M. Frank, *et al. IEEE Trans. Nucl. Sci.,* **NS–12, No. 5, 126** (1965).

252. A.D. Rossin in: Dosimetry for radiation damage studies. Report ANL-6826, March 1964; *IEEE Trans. Nucl. Sci.,* **NS–11**, No.5, (1964).

253. H. Stein, *J. Appl. Phys. Letters,* 2, 235 (1963).

254. J.M. Kortright, *et al. Bull, Amer. Phys. Soc.,* 7, 330 (1962).

255. E.E. Klontz, *et al. J. Phys. Soc. Japan,* **18**, III, 216 (1963).

256. S. Ishino *et al.* Report on Symposium on Lattice Defects in Semiconductors. Tokyo, Japan, 1966.

257. R. Genre *et al.* Report on International Conf. on Radiation Effects in Semiconductors. Toulouse, March 1967.

258. C.A. Goben, *IEEE Trans. Nucl. Sci.,* **NS–12**, No. 5,p.134 (1965).

259. N.G. Dobrokhotov, Semiconductor Devices and their Application (in Russian) Issue 7, Edited by D.A. Fedotov, *Sovetskoe Radio,* 1961, 3.

260. V. Pasynkov, *et al.* Semiconductor Devices (in Russian) Moscow, Vysshaya Shkola, 1966, 306.

261. F.A. Leith, *et al. IEEE Trans. Nucl. Sci.,* **NS–12**, No. 6, 64 (1965).

262. M. Frank, *et al. IEEE Trans. Nucl. Sci.,* **NS–10, No. 5, 93 (1963).**

263. V.A. Stafeev, *Fizika Tverdogo Tela,* 1, 848 (1959).

264. S.K. Manlief, *IEEE Trans. Nucl. Sci.,* **NS–11**, No. 5, 47 (1964).

265. J.M. Shwartz, *et al. J. Appl. Phys.,* 37, 745 (1966).

266. D. Billington, J. Crawford, Radiation Damage in Solids, Princeton, 1961.

267. J. Blackmore, Electron Statistics in Semiconductors, (in Russian) Moscow, Mir, 1964.

268. H. James, Semiconductor Materials, (in Russian) Edited by V.M. Tuchkevich, Moscow, Izd-vo Inostr. Lit., 1954.

269. R.F. Konopleva, et al. Report to the International Conf. on Radiation Effects in Semiconductors, Toulouse, March 1967.

270, I.P. Akimehenko et al. Fizika Tverdogo Tela, (1964).

271. Yu. A. Kazanskii et al. Physical Investigations of Reactor Shielding (in Russian), Moscow, Atomizdat, 1966.

272. Ya.A. Fedotov, Physical Principles of Semiconductor Devices, (in Russian), Moscow, Sovetskoe Radio, 1963.

273. S.D. Dodik, Semiconductor Stabilizers of Direct Voltage and Current (in Russian), Moscow, Sovetskoe Radio, 1962.

274. E.A. Carr, IEEE Trans. Nucl. Sci. NS–11, No. 5, 12 (1964); NS–12, No. 5, 30 (1965).

275. W.A. Bohan, et al. Proc. IEE, 106, Part B, Supplement No. 15, 361 (1959).

276. H. Rzewuski, In: Proceedings of the 7th International Conf. on the Physics of Semiconductors, Part 3, Paris, 1965, p. 35.

277. Influence of Radiation on Materials and Devices in Electronic Circuits, (in Russian), Translated from English by V.N. Bykov and S.P. Solov'eva, Moscow, Atomizdat, 1967, 311.

278. R.M. Kloepper, IEEE Trans. Nucl. Sci., NS–11, No. 5, 137 (1964).

279. H.J. Stein, et al. Trans. Radiation Defects. Report on Conf. on Irradiation Effects in Semiconductors. Toulouse, 7–11 March, 1967.

280. H.H. Sander, IEEE Conf. on Nucl. Radiation Effects, Seattle, Washington, July 1964.

281. H.H. Sander et al. Trans. IEEE Nucl. Sci., NS–13, 53 (1966).

282. D. Binder et al. Report on IEEE Conf. on Nuclear Space Radiation Effects. Palo Alto, California, July 1966.

283. G.A. Gilmour, Trans. IEEE Nucl. Sci., NS–10, No. 4, 18 (1963).

284. V.A. Ditkin, P.I. Kuznetsov, Handbook on Operational Methods in Computing (in Russian), Moscow-Leningrad, Gostekh-

teorizdat, 1951.

285. J.L. Wirth, *et al. Trans. IEEE Nucl. Sci.*, **NS—11**, No. 5, 24 (1964).

286. R. Beaufoy, *et al. A.T.E. Journal*, **13**, 310 (1957).

287. J.C.Henderson, *et al. Proceedings of the IEEE*, 104, Part B, No. 15, 318 (1957).

288. Yu.P. Nosov, Semiconductor Pulsed Diodes, (in Russian), Moscow, Sovetskoe Radio, 1965.

289. C. Rosenberg, *et al. Trans. IEEE Nucl. Sci.*, **NS—10**, No. 5 149 (1963).

290. R.S. Caldwell *et al. Commun. and Electronics.* No. 64, 483 (1963).

291. J.J. Loferski, *J. Appl. Phys.*, **29**, 35 (1958).

292. J.W. Easley, *IRE Weskon Conv. Rec.*, 3, 149 (1958).

293. R.R. Blair *et al.* Proceedings of the Second Conf. on Nuclear Radiation Effects on Semiconductor Devices. Materials and Circuits, Cowan Publish. Corp. N.Y., 1959, p. 96.

294. K. Zander, *Nucleonik*, 3, No.7, p.292 (1961).

295. L.B. Garner, *et al. IRE Trans. Nucl. Sci.*, **NS—8**, No. 3, 35 (1961).

296. F.W. Poblenz, *IEEE Trans.Nucl. Sci.*, **No. 5**, No. 1, 74 (1963).

297. D.S. Peck, *et al. Bell System Techn. J.*, **42**, 05 (1963).

298. R.S. Caldwell In: Inst. Environmental Sci. Annual Meet. Proc., Los Angeles, Calif., 1963, Mt. Prospect, 1963, p. 145.

299. S.C. Rogers, *Ibid.*, p. 270.

300. G.J. Brucker *et al. IEEE Trans. Nucl. Sci.*, **NS—12**, No. 5, 69 (1965).

301. D.S. Gage, *IEEE Trans. Nucl. Sci.*, **NS—12**, No. 5, 112 (1965).

302. P.A. Andrews, *Proc. IEEE*, **53**, 1653 (1965).

303. G.J. Brucker, *Ibid.*, **53**, 1800 (1965)

304. H.L. Hughes, *IEEE Trans. Nucl. Sci.*, **NS—12**, No. 6, 53 (1965).

305. J.B. Compton, *et al. IEEE Trans. Aerospace*, 3, No. 2, 378 (1965).

306. Ph. Glotin, *Bull. Inform. Sci. Techn.*, No. 98, 69 (1965).

307. Shao Bin-Sen *et al. Acta Sci. Natur. Univers. Fudan*, **10**, No. 2, 281 (1965).

308. W. Rosenzweig, *IEEE Trans. Nucl. Sci.*, **NS–12**, No.5, 18 (1965).

309. S.C. Rogers, *IEEE Intern. Conf. Rec.*, **14**, No. 10, 136 (1966).

310. W.E. Horne, *IEEE Trans. Nucl. Sci.*, **NS–13**, No. 6, 181 (1966).

311. J.P. Mitchell, *et al. Bell System Techn. J.*, **46**, No.1, 1 (1967).

312. J. Raymond, *IEEE Trans. Nucl. Sci.*, **NS–12**, No. 5, 55 (1965).